Water Vole Field Signs and Habitat Assessment

Water Vole Field Signs and Habitat Assessment

A PRACTICAL GUIDE TO WATER VOLE SURVEYS

Mike Dean

Pelagic Publishing | www.pelagicpublishing.com

Published by Pelagic Publishing
PO Box 874
Exeter
EX3 9BR
UK

www.pelagicpublishing.com

Water Vole Field Signs and Habitat Assessment:
A Practical Guide to Water Vole Surveys

ISBN 978-1-78427-254-8 (Pbk)
ISBN 978-1-78427-255-5 (ePub)
ISBN 978-1-78427-256-2 (ePDF)

A CIP record for this book is available from the British Library

All photographs © Mike Dean unless otherwise stated.

Cover images: Main image: Water vole © Simon Booth Photography.
Other images (left to right): Optimal water vole habitat provided
by the River Coln in Gloucestershire; water vole feeding
remains; water vole burrow. All photos © Mike Dean.

Back cover images: Optimal water vole habitat provided by a
drainage ditch in Kent © Mike Dean; water vole latrine © Mike Dean;
water vole swimming © Jo Cartmell.

For Xander and Jessica

Contents

Acknowledgements

I must start by thanking Merryl Gelling and Robyn Stewart, both experienced ecologists with considerable expertise on water voles. They were kind enough to review an early draft of this book, and it is much improved as a result of their insightful comments.

This was always intended as a pictorial guide, and it is necessarily reliant on good-quality images. I am therefore incredibly grateful to all those who have kindly allowed me to use their photos in this book: Simon Booth (Simon Booth Photography), Jo Cartmell (NearbyWild), Coral Edgcumbe, Derek Gow, Gareth Harris, Magnus Johnson, Kevin O'Hara, Stefanie Scott and Robyn Stewart. Thank you also to Hugh Brazier, who skilfully edited my initial text, and to BBR Design who have patiently turned the text and the jumble of photos that I sent into this book.

I am forever indebted to Rob Strachan. Without his knowledge of water voles, his willingness to share this knowledge, and his passion for their conservation, I wouldn't have been able to write this book. I'm therefore delighted to have been granted permission to reproduce his drawings of water vole and rat footprints, taken from his book *Water Voles*, and I'm grateful to Jane Sedgeley-Strachan and Whittet Books for giving me consent to do so.

1. Introduction

A bit of background

As I sit down to write this book, the water vole (*Arvicola amphibius*, previously *A. terrestris*) has been a protected species in the United Kingdom for more than 20 years, and has been identified as one of the UK's most rapidly declining mammals for over three decades. And it's probably fair to say that over those last 20–30 years there's been a significant amount written about this species. What else, you may reasonably ask, is there left to say?

The level of attention that the water vole has received in recent times also means that many professional ecologists and amateur naturalists will have had the opportunity to undertake, or get involved in, a water vole survey. They may have recorded its field signs or even seen the animal itself. Why then, do we need another book on this species?

Well, the answer is this: water voles are still declining and the number of water voles in the UK is predicted to continue falling (Mammal Society 2018). There is an increasing need for action to conserve them, and an urgent necessity for them to be considered appropriately in the context of development projects. And this means that surveys must not only be carried out, but done well. Over the years that I've been working on water voles, I've come across many situations where this hasn't been the case – where surveys have failed to spot their presence, perhaps because the animals were using habitat that the surveyor assumed they wouldn't, or because the field signs were interpreted incorrectly. This can be costly for a developer, if the failure occurs during a survey to inform a planning application, for example, or could even result in a prosecution if the mistake isn't spotted in time. On the other hand, in some cases, it may never come to light. But irrespective of the scenario in which it happens, or the reasons for it, the main 'loser' is the water vole. Important protection measures may not be put in place, appropriate management activities may not be undertaken, and conservation efforts may not be directed to the correct locations.

This book, then, is my attempt to help minimise the likelihood of such errors happening in the future.

Firstly, it's critical that those individuals involved in undertaking water vole surveys, or ecologists tasked with determining where to survey for water voles, or deciding whether a water vole survey is needed, have a good understanding of the range of habitats that this species will use. What are its preferred habitats? But also, where might you still find it, even where the habitat is less than optimal? So, in this book you will find a detailed description of some of the different habitats that water voles use, illustrated with photographs. I've included an explanation of how water voles use the habitat, thereby allowing the surveyor to assess how suitable or otherwise a given habitat is, and also to apply the principles to specific habitat types that I haven't covered.

Secondly, when we undertake water vole surveys we are searching for field signs. These provide us with valuable information, provided that we can find them, and that we can identify them correctly. There have been many occasions when I know that surveyors have failed to do this. So what you'll also find in this book is a very detailed description of water vole field signs, illustrated with lots of photographs (who doesn't like a photo of some poo?), as well as tips on how to distinguish these from those of other similar species (yes, you guessed it, you may well have to sniff it!).

This book is aimed at professional ecologists and amateur naturalists. In fact, it's aimed at anyone doing a water vole survey for any reason at all.

What I've specifically *not* done in this book is give advice on how to design a good water vole survey, how to interpret the results of those surveys, how to manage habitat for water voles, or how to assess impacts of development projects on water voles or design mitigation for those impacts. I've also not provided a detailed description of the ecology of the species, their conservation status or their UK distribution. These topics are all covered in detail already in other publications, particularly the Mammal Society's *The Water Vole Mitigation Handbook* (Dean *et al.* 2016) and the *Water Vole Conservation Handbook* published by Oxford University's Wildlife Conservation Research Unit (Strachan *et al.* 2011). This book aims to supplement those handbooks, rather than replace them.

There are a number of other reference sources that you can look at, and I would encourage you to do so; I've listed some of these in the Bibliography. I've made reference to some of these sources in this book, as evidence for some of my statements or recommendations. However, the majority of the information provided is based on my own personal experience, rather than peer-reviewed publications, which leads me neatly on to …

Personal experience

Any ecologist who ends up spending an inordinate amount of their time working with one particular species will have a story to tell as to how that came about, and who helped them on their way. Here is mine.

I started doing surveys for water voles in early 1998, when the species first became protected. I'd started working for an ecological consultancy in Nottingham, and was tasked with finalising the ecological assessment work for a new pipeline, to be laid between Nottingham and Derby. This pipeline crossed a number of slow-flowing watercourses and canals – prime water vole habitat. With the inclusion of water voles on Schedule 5 of the Wildlife and Countryside Act (see below), and the identification of their decline in the early 1990s leading to them being considered in the UK Biodiversity Action Plan as a 'Priority Species' for conservation (a term now largely superseded by 'Species of Principal Importance for the Conservation of Biodiversity'), there was a sudden need to undertake surveys, assess impacts and design mitigation. The problem was that very few people knew how to survey for water voles, and nobody had a good idea of how they might be affected by a development, or how to mitigate for those impacts.

It was fortunate for me that the one person who knew more than anyone else about this species, the late Rob Strachan, was willing to share his knowledge and expertise. The next few years were a steep learning curve for me, as well as for other ecologists trying desperately to work out the answers as to how we conserve water voles. I coordinated a county-wide survey of Nottinghamshire for water voles, using volunteers. And I was involved in two national conferences on water voles aimed at sharing good practice on conservation and mitigation measures, both held in the late 1990s.

Figure 1.1　Water vole feeding on watercress. © Jo Cartmell.

Figure 1.2   Infant water vole feeding on reed sweet-grass (*Glyceria maxima*). © Jo Cartmell.

Since 1998 I've undertaken a lot of water vole surveys (including surveys to target conservation measures appropriately and those required in relation to construction projects), assessed the impacts of numerous proposed developments on this species, designed and implemented mitigation measures at countless sites, carried out research into the effectiveness of water vole mitigation (Dean 2003, Gelling *et al.* 2018), trained many ecologists in surveying for water voles and designing mitigation, and co-authored *The Water Vole Mitigation Handbook*, which was published by the Mammal Society (Dean *et al.* 2016). I realise that the previous sentence contains a series of vague terms, such as 'a lot', 'numerous', 'countless' and 'many', but I don't know actual numbers I'm afraid, and I doubt if any others working on water voles do either, so the numbers would be meaningless even if I could provide them. Most of my experience relates to water voles in lowland England and so they are the focus of this book, although I've tried to cover other scenarios as well.

Legal protection in the UK

It's important that I say something about the legal protection afforded to water voles. There may be legislation relating to water voles in other parts of the world, but I am dealing only with the UK situation here. It will be up to any individual undertaking a survey or carrying out other activities that might affect water voles, such as water-course maintenance, to ensure that they are conversant with the most up-to-date legislation, and the need for their activities to be carried out under a licence issued by the relevant Statutory Nature Conservation Body.

At the time of writing, water voles in England and Wales are protected under the Wildlife and Countryside Act 1981 (as amended). They were initially given partial protection, in 1998, but this was extended to full protection in 2008. The Act, together with amending legislation, makes it an offence to:

- Intentionally kill, injure or take a water vole.
- Possess or control any live or dead water vole, or any part or derivative.
- Intentionally or recklessly damage or destroy a structure or place used by a water vole for shelter or protection.
- Intentionally or recklessly disturb a water vole whilst it is occupying a structure or place used for shelter or protection.
- Intentionally or recklessly obstruct access to a structure or place used by a water vole for shelter or protection.
- Sell, offer for sale, or possess or transport for the purpose of sale, any live or dead water vole, or any part or derivative, or advertise any of these for buying or selling.

In Scotland, at the time of writing, water voles receive partial protection under the Wildlife and Countryside Act 1981 (as amended), making it an offence to:

- Intentionally or recklessly damage or destroy a structure or place used by a water vole for shelter or protection.
- Intentionally or recklessly disturb a water vole whilst it is occupying a structure or place used for shelter or protection.
- Intentionally or recklessly obstruct access to a structure or place used by a water vole for shelter or protection.

Competence

It's clearly vital that those people undertaking water vole surveys, for whatever purpose, are competent to do so. They need to know what they're looking for, where to look for it, when to look for it, and how to look for it. They need to understand the limitations of different survey methods, and they need to be able to acknowledge their own personal limitations. This book aims to help with some of these issues – specifically what to look for and where to look for it. However, it should not be seen as sufficient on its own. You will not become a competent water vole surveyor just by reading this book. You will need to gain practical experience and be familiar with the handbooks that I've referred to already, as well as many of the other reference sources listed in the Bibliography.

Once you are competent – and hopefully the following chapters will help you on the way if you aren't there already – this book should provide a useful reference source for when you find yourself assessing a habitat, or identifying a field sign, that you're not sure about.

2. Habitat

Where do you find water voles?

In continental Europe water voles are found in a range of terrestrial habitats, including meadows and pastures. In the UK, however, water voles are normally associated with wetland habitat, although there are exceptions, as I'll come on to later.

So, for the most part, we can say that water voles in the UK occur in wetland systems, including both flowing and static waterbodies, such as rivers, streams, canals, ponds, lakes, lagoons, reedbeds, drainage ditches and fens.

What does 'ideal' habitat for water voles look like?

The national surveys undertaken by the Vincent Wildlife Trust in 1989–1990, and again in 1996–1998, give us a picture of the variety of habitats that water voles occur in (Strachan and Jefferies 1993, Strachan *et al.* 2000). These studies collected data on the characteristics of the habitat, and allow us to pick out those which water voles exhibit a preference for. Based on these studies, the 'ideal' or 'optimal' water vole habitat would be characterised as a slow-flowing watercourse, less than 3m wide, around 1m deep, with limited fluctuation of water levels, steep earth banks, a lack of shading by trees or scrub, and a continuous swathe of tall and luxurious riparian vegetation providing at least 60% ground cover. Further information can be found in Rob Strachan's *Water Voles* (1997).

There's no shortage of rivers and streams that meet these criteria near where I live in the Cotswolds. These are in the upper reaches of the catchment of the River Thames, and therefore are not hugely prone to flooding. They generally look something a little bit like the stretch of the River Coln shown in Figure 2.1. Water vole habitat doesn't get much better than this. However, water voles occur in a wide variety of different habitats that might also be considered 'optimal', even where they don't quite so obviously meet the criteria described above.

I've labelled the habitat characteristics that I think are important for water voles on Figure 2.1. And I've provided a series of photos of habitat that might also be considered 'optimal' in Figures 2.2 to 2.10. You could move the 'habitat characteristics' labels from Figure 2.1 onto each of those figures, with relatively few changes.

Very few bankside trees and
therefore little shading

Steep earth banks
suitable for burrowing

Slow-flowing river, with little
fluctuation in water levels

Continuous cover of
riparian vegetation

Figure 2.1   The River Coln in Gloucestershire – optimal water vole habitat.

The obvious characteristic that all of these situations have in common is that the bankside and in-channel habitat is relatively unshaded, allowing the development of continuous cover of luxuriant riparian vegetation and tussocky bankside grasses. I'll come on to the reason that this is so important later.

The other characteristic they all share is that the water levels will fluctuate little, if at all. Some are slow-flowing rivers. Others are drainage ditches in situations where the levels are unlikely to change dramatically on a daily, or even monthly, basis. And others are static waterbodies (a canal, a pond and a lagoon) where the levels are likely to be fairly constant. I've also included an example of a system of drainage ditches where the water levels are controlled through a series of sluices, and therefore have a 'summer level' and a 'winter level', with little variation in between.

In my experience the width and depth of the watercourse or waterbody is relatively unimportant. You'll notice that I haven't labelled the width and depth of the river in Figure 2.1. Water voles occur on very wide watercourses, such as main rivers and canals where the banks are tens of metres apart, as well as on very narrow watercourses, including many that are less than a metre wide. And consequently they'll occur on very shallow watercourses or waterbodies, as well as very deep ones.

In most of the example photos I've provided, the habitat includes a steep earth bank for water voles to burrow into, which is immediately adjacent to the water's edge. This is also a defining characteristic of 'optimal' water vole habitat in most cases, as water voles live in burrow systems and need these to be close to water. However, water voles also live in above-ground woven nests, using these particularly where there is no bank, or the bank is too far from the water's edge. Figure 2.3, of a pond in an urban setting, highlights this – as I've labelled on that photo. And Figure 2.6 also shows an extensive bed of reed sweet-grass (*Glyceria maxima*) on one side of a river – this is also the sort of situation in which water voles will construct above-ground woven nests. More about these in Chapter 8.

I've included two examples of recently constructed habitat in Figures 2.7 and 2.8. The lagoon in Figure 2.7 was colonised by water voles within approximately three years of it being finished. Figure 2.8 shows the extension of an existing ditch that supported water voles – the new section was colonised in less than twelve months of being constructed. Both examples constitute 'optimal' habitat.

There are some extensive water vole populations associated with upland areas. Typically the animals use small streams or boggy areas. Some of the streams may be so narrow (less than 15 cm wide) and so deeply cut into the peat, that it's difficult to view the 'toe' of the bank where evidence of water voles is likely to be found (Figure 2.9). In other cases they use wetland habitat that pools across the surface of the peat, with little in the way of obvious banks at all.

Figure 2.2   'Optimal' water vole habitat along a disused section of canal.

Bank is shallow-sloping; suitable areas for burrows are set well back from the water's edge

Very wide fringe of common reed is suitable to support above-ground nests

Figure 2.3 'Optimal' water vole habitat provided by a pond in an urban setting.

Figure 2.4 'Optimal' water vole habitat along a lowland drainage ditch.

Figure 2.5 Extensive network of drainage ditches in eastern England with controlled water levels, providing 'optimal' habitat.

Figure 2.6 Lowland river, with a very wide bed of reed sweet-grass suitable for above-ground nests, providing 'optimal' habitat.

Figure 2.7 Recently constructed attenuation lagoon, with extensive beds of common reed (*Phragmites australis*) as well as good marginal vegetation, providing 'optimal' habitat.

Figure 2.8 Newly created ditch providing optimal habitat for water voles. The marginal/bankside vegetation has been established by plug planting reed sweet-grass on the left bank and installing pre-grown coir pallets on the right bank.

Figure 2.9 Narrow stream in the uplands, overhung with rushes and grasses, providing 'optimal' water vole habitat. There is little emergent vegetation, but given the narrowness of the channel there is sufficient food and cover without it. © Kevin O'Hara.

Figure 2.10 Lowland river that provides excellent habitat for water voles (the same river shown in Figure 2.6) but taken when the river is in spate. Emergent vegetation is still visible, but the full extent of it is impossible to determine under these conditions. There is some 'headroom' left in the banks, which may allow water voles to survive until water levels recede; no burrow entrances are visible.

Habitat requirements

As many of those reading this book will already know, water voles turn up in a considerable number of places that seem to be a million miles away from the examples of 'optimal' water vole habitat that I've just given. The key question, then, is what do water voles actually need? What are the habitat characteristics that they require? Over the years of studying this species, I've come to the conclusion that there are three basic habitat characteristics, which are just an extension of what I've already shown with those 'ideal' or 'optimal' scenarios – we just have to think carefully about each of these, as there is a broad spectrum of habitats that meet the characteristics without being so obviously 'optimal'.

These habitat characteristics are driven by the two key aspects that enable water voles to survive – access to food, and avoidance of predators. It's worth remembering that water voles have a wide variety of predators – name a predatory mammal or bird and the chances are that water voles will form part of their diet where the two species coexist. In many environments, they will therefore tend to associate with habitats that provide them with opportunities to avoid predation. And water voles have three basic predator avoidance tactics:

1   Go underground into a burrow that a larger mammal, or bird, can't follow you into.
2   Sit tight, be quiet and still, and hope that you aren't seen.
3   Dive into the water, particularly to escape mammals that hunt on the banks but don't like getting their feet wet.

These then lead us on to our key habitat characteristics. In my experience, with the exception of where water voles are using terrestrial habitat (more about this later), the species needs three basic things in relatively close proximity:

1   Dry areas for burrows, or above-ground nests where burrowing is not possible.
2   Herbaceous vegetation as food and cover.
3   Water.

These three characteristics are also identified in *The Water Vole Mitigation Handbook* (Dean *et al.* 2016).

So, when you start to consider whether a habitat is suitable to support water voles, these are the three factors to consider. Let's look at each in turn.

Dry areas for burrows or above-ground nests

Why are slow-flowing watercourses preferred over fast-flowing ones? Is it that water voles aren't actually very good swimmers? Maybe that's part of it, but I don't think it's a major factor. In my view the answer probably comes down to three things. Firstly, fast-flowing rivers tend to have less in-channel emergent vegetation, which will restrict the amount of food and cover for water voles (see *Herbaceous vegetation as food and cover*, below), as well as removing their ability to create above-ground nests. Secondly, fast-flowing rivers are more likely to have rocky or stony banks, rather than earth banks, which make burrowing difficult (not impossible, as we'll see, but certainly more difficult). And thirdly, fast-flowing rivers will tend to have banks with a shallower profile, and the water levels will vary more than they do on slower-flowing rivers.

You might argue that I've combined two separate elements into my third point – bank profile and variation in water level. And you'd be right, but I've done that deliberately as I see the two things as related to each other. Water voles prefer steep banks for burrowing. However, they will create burrows in banks with a very shallow profile. This tends to occur on watercourses with very little variation in water level or, more often, static waterbodies with no real variation in levels. Where a watercourse has both a bank with a shallow profile and a considerable variation in water level it will be difficult for water voles to excavate a burrow system that enables them to respond to such changes. It would have to be very extensive.

Of course, if the bank of a watercourse becomes completely submerged during a flood event (so that the levels rise to effectively 'overtop' the bank) then this will result in the temporary loss of any 'dry areas' – for burrows, at least, and probably also for above-ground nests. Water vole populations do appear to be capable of surviving in cases where this occurs relatively infrequently. It might be that the individual animals resident at the time of the flood move away from the banks, find somewhere safe and dry, and return when the levels drop. It might be that the resident animals simply don't survive and the habitat is recolonised from areas that were less badly affected. Or perhaps it's a combination of the two. Water vole populations are less likely to be found on watercourses where the banks overtop very regularly. On the other hand, water voles only require a few centimetres of dry bank above high-water levels to survive in – see Figure 2.10 for example.

So, when you consider whether a particular wetland habitat provides dry areas for burrows or nests for water voles, you need to understand what the variation in water levels is likely to be – both the normal range of fluctuation and the flood levels (as well the frequency of flooding). This might be obvious from the positioning of flood debris, for instance, but it won't always be, and this is one of the reasons that *The Water Vole Mitigation Handbook* (Dean *et al.* 2016) recommends more than one visit to a habitat to survey it for water voles – in order to detect such changes. A desk-based review will also provide useful information on flood risk.

You will also need to consider the bank substrate. Is it suitable for water voles to burrow into? This isn't always that simple. Water voles obviously prefer an earth bank, as a burrow can be excavated relatively quickly and with minimal likelihood of it collapsing. Banks that are very sandy will be much more difficult for water voles to create burrows in, although you will find water vole burrows in banks that are a mixture of earth, sand and gravel (see Figure 2.11). Water voles will also excavate burrows in peat, and there are numerous examples of water voles doing so in upland habitats across the UK. Water voles may also use banks that are part stony and part earth – see Figures 2.12 and 2.13 for instance.

Water voles are also pretty adept at creating burrows in reinforced banks. They are found in stone-lined channels, where there are gaps in the stonework allowing the animals access to the earth bank behind (Figures 2.14 and 2.15), and even in channels with mortared concrete or sheet piling reinforcement, where the animals can climb up and over this, excavating burrows in the bank directly behind. Of course, there will come a point where the level of reinforcement is such that water voles cannot use the habitat, but the point is that watercourses like this should not be automatically discounted. And there will be cases where one bank may be made unsuitable for burrowing by extensive sheet piling, for example, but where the other bank provides suitable habitat; water voles do occur in such situations (Figure 2.16).

Herbaceous vegetation as food and cover

Water voles are largely herbivorous, and so they need vegetation as food. It also provides them with cover to avoid predation.

Wide fringes of luxuriant marginal vegetation, such as reed sweet-grass, branched bur-reed (*Sparganium erectum*) or yellow iris (*Iris pseudacorus*), provide preferred foods as well as excellent cover. This means that water voles can find a spot that is completely surrounded by the vegetation that they want to eat, making it ideal. Research has shown that the wider the fringe of marginal vegetation the higher the density of water voles (Moorhouse *et al.* 2009). We can therefore judge habitat suitability, in part, based on the extent of such vegetation.

However, water voles will eat a wide variety of plants, not just the luxuriant wetland vegetation that is present in those 'optimal' habitats. They will survive on watercourses where the banks are dominated by common nettle (*Urtica dioica*) or great willowherb (*Epilobium hirsutum*), for example, and there is little or no marginal vegetation (Figures 2.11, 2.12, 2.17 and 2.18).

A few years ago I watched a water vole, on a river close to where I live, as it fed on the only bit of vegetation available to it – sycamore (*Acer pseudoplatanus*) leaves (Figure 2.14). It obtained these by felling a very small sycamore sapling growing at the base of the bank, and then eating each leaf in turn. Water voles have also been known to strip bark off trees and eat berries during autumn and winter. So, whilst it might be fairly straightforward to identify those habitats that are 'optimal' in terms of their vegetation, it is much more difficult to rule any out on this basis.

Water voles also occur in situations where the vegetation provides little in the way of cover. They will occur on watercourses that have short, regularly mown grass on the banks (Figures 2.16, 2.19 and 2.20). This isn't going to be ideal for water voles, but that doesn't mean that they won't occur there.

In the late 1990s I remember running a training course for people volunteering to help with a water vole survey of Nottinghamshire. We walked alongside a stretch of densely vegetated drainage ditch in the centre of Nottingham, known as the Tinker's Leen. I waxed lyrical about the high quality of this habitat in comparison to the poor habitat provided by the canal that ran immediately adjacent and parallel to it. And just as I was doing so, we all watched as a water vole swam across the canal, dodging round the front of a barge that was making its way past us, climbed up a low section of sheet piling on the far side, and settled down in an area of short grass to graze alongside several ducks. Quite clearly, this particular urban water vole had not read the textbooks!

I've found water voles in this sort of habitat on numerous occasions since. Water voles seem to be more often found in habitats with limited cover in urban or suburban situations – perhaps because there is less predation pressure. The banks of water-courses in urban or suburban environments are often cut very frequently and, whilst we can't rule water voles out in such situations, this will put them under increased pressure. Cutting the banks infrequently, if feasible given the drainage function of the watercourses, is likely to be beneficial for water voles.

When assessing the vegetation associated with a watercourse, it is particularly important to understand how it is managed, how frequently this occurs, and whether any management operations have been undertaken in the months immediately prior to the survey, as this can obviously influence the assessment (Figures 2.20 and 2.21).

Water

Water voles, in the UK, are found close to water. Very close to water – basically within a metre or two of it. There are exceptions, as I'll come to in a while. But, in the vast majority of cases, water voles want to excavate their burrow system or create their above-ground nest, and have access to herbaceous vegetation for food and cover, within 2m of water. The burrow systems can extend further than this from the water's edge, so you will find the animals slightly further away as well. But if the dry areas and herbaceous vegetation identified as the two previous habitat characteristics are not within 2m of water, it's significantly less likely that the habitat will be occupied by water voles.

Why do water voles associate with water? One theory is that it is a predator avoidance measure. A significant number of our native mammals struggle to hunt in wetland habitat – stoats (*Mustela erminea*), weasels (*M. nivalis*) and foxes (*Vulpes vulpes*), for example. The same applies to domestic cats, which can also prey on water voles. So having quick access to water, where they can escape from some (but not all) predators, is thought to be important for water voles.

Another reason for the association of water voles with water might be the fact that watercourses, as a general rule, tend to provide the open, less intensively managed habitat that supports the herbaceous vegetation that water voles want as food and cover. And particularly the more luxuriant vegetation that is typically associated with riparian habitats. However, when a ditch dries out permanently you tend to find that the water voles move away into surrounding wet habitats, even if the vegetation is essentially unchanged, suggesting that the presence of water itself is key. There are, though, instances where water voles will remain in dry ditches, particularly if the drying out is temporary or where the water vole population is at very high density – meaning that there's nowhere else for the animals to move to.

How to assess likely value of habitat

Assessing the likely value of habitat for water voles, as with any other species, is quite subjective. What one person considers suitable, or good, or amazing, can be very different from what another person might pick out. It's difficult to get away from subjectivity completely, as the range of different habitats that water voles use varies widely and we'll all have different experiences of where we've found them. Coupled with this, there is considerable variation within a habitat type, in terms of the specific characteristics that are favourable for water voles. So some 'professional judgement' is required. (Professional judgement is defined in the British Standard on *Biodiversity – Code of Practice for Planning and Development*, BS42020:2013, as 'use of accumulated knowledge and experience in order to make an informed decision that is clearly capable of being substantiated with supporting evidence'.) However, in Table 2.1 I've attempted to piece together the variables in a way that allows a greater degree of consistency when I use terms to describe the quality of a habitat for water voles.

The next series of photographs (Figures 2.11 to 2.23) show a variety of habitats that I would classify as suitable for water voles, but would not consider to be 'optimal'. In each case, I've indicated whether I would consider the habitat to be 'good' or 'suitable but poor', and I have tried to explain the reasons for my assessment. Note that water voles were present at all of these sites when the photos were taken.

Figure 2.11 Mill stream linked to a main river which supports water voles. One bank is heavily shaded and the other has a stony substrate, making it difficult for burrowing. There is also a general lack of luxurious emergent vegetation. I would assess this as 'suitable but poor' habitat; water voles are present and there are small numbers of burrows on both banks.

Table 2.1 Assessing the value of habitat for water voles (excluding terrestrial/fossorial populations)

Habitat category	Dry areas for burrows or nests			Herbaceous vegetation	Water
	Bank profile	**Bank substrate**	**Variation in water level**		
Optimal (all criteria need to be met)	Steep (approaching 1:1) on at least one side of a watercourse. Steep or shallow banks on static waterbodies or fen-type habitat, where water levels do not fluctuate significantly	Earth or peat	No noticeable variation during the summer months; banks are not overtopped regularly [a]	Continuous swathe of tall and luxurious riparian vegetation providing 90–100% cover on the banks (tall tussocky grassland) and marginal/in-channel vegetation is present (emergent species)	Permanent water
Good (all criteria need to be met)	Steep (approaching 1:1) on at least one side of a watercourse. Steep or shallow banks on static waterbodies or fen-type habitat, where water levels do not fluctuate significantly	Earth or peat banks, or stony/reinforced bank with gaps allowing access to the earth behind	No noticeable variation during the summer months; banks are not overtopped regularly	Continuous swathe of bankside or in-channel (emergent) vegetation providing at least 60% ground cover. May be dominated by grasses and weeds, rather than luxurious riparian vegetation. The vegetation should generally be tall, except in urban or suburban areas, where shorter bankside vegetation may also qualify	Permanent water. Or routinely wet for at least 2–3 months during the summer, and where other 'good' habitat is present in immediately adjacent areas with permanent water

Suitable but poor [b]	Any habitat that falls short of the criteria to qualify as 'good' but does not meet the criteria of 'negligible value' could reasonably be considered to be 'suitable but poor'				n/a
Negligible value (will generally need to meet the criteria for herbaceous vegetation and at least one other)	Shallow profile on both banks	Rocky or gravel, unsuitable for burrowing	Considerable variation in water level – the bank toe can move by more than 1 m horizontally over the breeding season	No or limited bankside and marginal vegetation (due to shading or other 'permanent' factors – note that management can change and is often a 'temporary' factor)	n/a
	Vertical bank face with no burrowing opportunities behind it	Reinforced banks with no gaps	n/a		n/a

[a] I can't be specific, but overtopping once every 5–10 years is likely to be too frequent in most cases; overtopping less frequently than this may also be problematic for water voles.

[b] I've used the term 'suitable but poor' to avoid the possible misinterpretation of this habitat category had I used the term 'poor', which some would dismiss as 'unsuitable' – this is clearly not the case, as water voles utilise habitat that is of low quality and which is therefore still 'suitable'.

Figure 2.12 Upland stream supporting a water vole colony. The stream bed is stony, and the banks are relatively stony as well. However, there is sufficient earth bank for water voles to burrow into and there is a good cover of bankside vegetation including tussocky grasses and, in places, tall weeds. I would consider this to be 'good' habitat. © Kevin O'Hara.

Figure 2.13 Upland habitat supporting water voles. The bed of the stream is stony, as are the lower parts of the banks. Nevertheless, this still fulfils the habitat requirements for water voles and I would assess this as 'good' habitat. © Derek Gow.

Figure 2.14 A stone-lined channel where water voles use gaps in the stonework for burrows. The emergent vegetation in the centre of the channel provides food and cover during the summer months, but the animals are reliant on species such as common nettle and sycamore for the remainder of the year. I would assess this as 'good' habitat, but if you assessed it in winter or spring, when there's no emergent vegetation visible, it would be easy to incorrectly classify it as 'suitable but poor' or even of 'negligible value'.

Figure 2.15 Stone-lined bank of a river. Water voles use the gaps in the stonework and access the vegetation immediately behind the bank by creating vertical burrows. There is a general lack of emergent vegetation within the river, and the animals are dependent on the meadow area behind the bank for food. I would therefore assess this as 'suitable but poor' habitat.

Figure 2.16 Urban river, with a steep grassy earth bank (but little cover) on one side and sheet piling on the opposite bank. Water voles are present and construct burrows in the earth bank but do not utilise sections with sheet piling (it is far too tall for them to climb up and over). I would assess this as 'suitable but poor' habitat, as only one bank is available and there is a lack of emergent vegetation. The bankside vegetation is also short, although that is less of a limiting factor on an urban watercourse such as this.

Figure 2.17 Narrow (lowland) ditch with steep and undercut earth banks that are dominated by common nettle and other weed species. I would assess this as 'good' habitat, rather than 'optimal', given the lack of emergent vegetation and low species diversity on the banksides.

Figure 2.18 A narrow lowland stream with steep earth banks and overhanging tussocky bankside grasses, but little emergent vegetation. I'd assess this as 'good' habitat. This stream is located adjacent to a network of ditches which provide better habitat (see Figure 2.4).

Figure 2.19 Lowland river with a frequently mown grass bank on the left and a scrubby, shaded bank on the right. I've included this photo to highlight the importance of understanding context. I would assess what you can see in this photo as 'suitable but poor' habitat, but in reality it is a short section within a river that provides 'optimal' habitat in adjacent areas (see Figure 2.6). Water voles have burrows in the bank on the left (see Figure 7.1), and are often seen from the footbridge.

Figure 2.20 Roadside ditch that is frequently dredged and with grassy banks that are frequently mown. Water voles are present but I would assess this as 'suitable but poor' habitat, because the frequency of management is preventing the development of a significant amount of marginal or emergent vegetation and the grass on the banks is cut relatively short.

Figure 2.21 Roadside ditch with steep earth banks, one of which is shaded and the other supports grasses and weeds. Water levels are relatively stable. You might assess what you can see in this photo as 'good' habitat, although it is borderline 'suitable but poor' due to the extent of shading. This section of ditch is contiguous with the section shown in Figure 2.20; the banks are cut and the ditch dredged very frequently, further reducing its likely value. Had management works occurred immediately before I took the photo the habitat would more closely resemble that in Figure 2.20, and it would be more confidently assessed as 'suitable but poor'. Understanding the management is critically important.

Figure 2.22 A well-vegetated drainage ditch within a housing estate. Good management is important in sustaining a water vole population in such situations, where frequent mowing is likely to be detrimental. I would assess this as 'good' habitat – in a rural setting, with better habitat links, and where there was a more reliably stable water level it would probably qualify as 'optimal'.

Figure 2.23 Temporary construction site pond with shallow banks, dominated by reedmace (*Typha latifolia*). It could be argued, based on what you can see in this photo, that the habitat is 'good' or possibly even 'optimal'. However, the pond is shallow and regularly dries out in summer and is not well linked to other suitable habitat. I would therefore assess it as 'suitable but poor' habitat.

Determining that a particular habitat is 'unsuitable' is particularly difficult. I've included some photos as examples of habitats that I would consider to be as close to 'unsuitable' as you're likely to get (Figures 2.24 to 2.27). However, I've described these as being of 'negligible value' for water voles, rather than 'unsuitable', because:

1   Water voles may occur in habitat that looks similar to what's shown in these photos; and

2   The context is important in ruling out the likely presence of water voles. If water voles were present on different lengths of the same watercourses (or sections of bank of the same waterbody) then I'd be uncomfortable about ruling them out completely. If there was no evidence of water voles in areas of better quality habitat on the same watercourse/waterbody, or in the wider area, then there's very little chance of them being present.

The importance of context is demonstrated in Figure 2.24, which shows a section of a lagoon. The bank in the foreground is of 'negligible value' for water voles, but further along the same bank the habitat improves and does support water voles. I didn't find water voles at the sites shown in Figures 2.25 to 2.27 inclusive.

Bank is suitable for burrowing, vegetation on the bank face is intact and some marginal vegetation is present

Bank has been trampled and vegetation grazed – habitat in this location is of 'negligible value'

Figure 2.24 A lagoon with stable water levels. The habitat in the immediate foreground is of 'negligible value', as the vegetation has been grazed and the bank trampled by wildfowl. However, the habitat on other adjacent sections is better, possibly 'good', with intact vegetation and an earth bank suitable for burrowing (the low bank is not a limiting factor, as water levels are stable).

Figure 2.25 Stream with steep, open grassy banks that are heavily grazed and trampled by cattle. This might be considered to be 'suitable but poor' habitat due to the level of grazing/trampling. However, discussions with the landowner highlighted that water levels fluctuate, with the banks overtopped in winter each year, and I therefore assessed it as being of 'negligible value'. If livestock were fenced off the bank, and the water levels didn't fluctuate so much, this would be more likely to support water voles.

Figure 2.26 Canal with one heavily shaded bank and one bank formed from pre-cast concrete blocks with no gaps between, and no area between the blocks and the towpath for animals to create burrows. I would assess this as being of 'negligible value'.

Figure 2.27 Stream with heavily shaded banks over long lengths (several hundred metres), which I would assess as being of 'negligible value'. It should be noted that water voles can exist in habitat similar to this where the shaded length is shorter (tens of metres) and water voles are present in more open sections upstream or downstream.

Water voles in terrestrial habitat

As I've already mentioned, there are water voles in the UK that live nowhere near water. These 'terrestrial' or 'fossorial' animals live in habitats that provide them with vegetation as food and cover, and an area in which they can create burrows, such as open grassland – the first two of the three characteristics I discussed earlier in this chapter. However, they appear to be able to survive without the third (water).

This is not unusual for the species in general. In continental Europe there are numerous examples of water voles existing, and even thriving, in such habitats. In the UK, though, it is less common – at least as far as we know. Most of the known examples of terrestrial or fossorial water voles in the UK are in Scotland. There are several examples in urban or suburban areas around Glasgow, where they are sometimes referred to as 'urban grassland water voles' (Stewart *et al.* 2018, 2019), and they have also been recorded in a noise bund adjacent to the M8 motorway. There are other examples outside the Central Belt of Scotland, such as in some upland areas.

It is likely that these animals can use both terrestrial and wetland habitat, and are simply making use of what is available.

As we'll see in subsequent chapters, the field signs of terrestrial or fossorial water voles are the same as those of water voles in wetland habitat, although you will obviously need to think differently about how and where you're going to search for them. See Figures 2.28 to 2.29 for some surprising examples of the sorts of locations where water voles have been found.

Figure 2.28 Area of long grass in a park which supports a terrestrial, or fossorial, water vole population. © Robyn Stewart.

Figure 2.29 Earth noise bund, covered in short grass, adjacent to a motorway, which supports a terrestrial, or fossorial, water vole population. © Robyn Stewart.

3. What does a water vole look like?

Water voles are generally either chestnut-brown in colour (Figures 3.1 and 3.3) or black (Figure 3.2). The brown form occurs throughout the UK and is slightly larger than the black form, which occurs mainly in Scotland. Both brown and black forms have been recorded in terrestrial or fossorial populations (Stewart *et al.* 2019).

When you see a water vole, the fur that is visible is fairly uniform in colour, although they are lighter on their undersides. The brown form sometimes has a white dot on its forehead, a white tail tip, or a white paw (or paws). The black form sometimes has flecks of white in its fur.

Water voles are a handful – both in terms of behaviour when being handled and in terms of size. The measurements of water voles quoted in the literature vary, and the black form is generally smaller than the brown form. In general though, the weight range for adult water voles starts somewhere between 100 and 150g and goes up to somewhere around 300–350g. Adults are approximately 14–22cm long (including head and body but excluding the tail). The average-sized water vole, excluding the tail, is therefore around the length of an adult human hand (wrist to tip of index finger). Of course hand sizes vary, as do water vole sizes, but other species of vole found in the UK are significantly smaller and will fit comfortably within an adult hand (see below).

Water voles have a blunt face and are generally very rounded in appearance. It can be difficult to locate the animal's neck when you see a water vole, particularly when it is sitting in the typical hunched posture (Figures 3.1 and 3.3). They have small eyes, and their ears do not protrude significantly beyond their fur, also making them difficult to discern in certain circumstances.

A water vole's tail is long in comparison with that of other UK vole species, being clearly more than half the length of its head and body. It is dark in colour, as it is covered with hairs the same colour as those on the body. The tail is visible in Figures 3.2 and 3.3.

Water voles tend to swim with much of their head and body high out of the water (Figure 3.4). They are capable of diving and swimming underwater, but can only do so for a few seconds. If you disturb a water vole, in clear water, you might see it dive to the bed of a stream, kick up some silt from the bottom, and then swim to dense vegetation or a burrow. They tend not to swim for long distances.

Brown rats (*Rattus norvegicus*) occur in similar habitats to water voles and are of a similar size and, superficially, similar appearance. In general, rats are larger, but there is some size overlap between a small rat and a large water vole. Rats have a more pointed face, more obviously visible ears, and a pink hairless tail (which is proportionately longer than a water vole's, when measured against head and body length). These are the key characteristics to look for if you see an animal of the right size. Rats also have greater variation in colouration between the fur on their backs (grey-brown) and undersides (pale grey).

Other species of vole in the UK are significantly smaller than water voles. They would fit easily within the palm of the hand. Bank voles (*Myodes glareolus*) have a head and body length of between 8 cm and 11 cm and an average weight of approximately 26 g (Harris and Yalden 2008). Field voles (*Microtus agrestis*) are slightly larger, with an average head and body length of approximately 10–12 cm and an adult weight of 30–40 g, though the largest male can weigh up to 55 g (Harris and Yalden 2008). The smallest water voles that you're likely to see above ground are around 50–70 g (brown form) or 40–45 g (black form). There is, therefore, a slight overlap in size between the largest field vole and the smallest water vole, but the average field vole is around 20% of the weight of the average water vole.

Field voles and bank voles also have tails that are significantly shorter than a water vole's. A field vole's tail is about one-third the length of the head and body; a bank vole's tail is proportionately slightly longer, but still less than half the length of the head and body.

Figure 3.1 Water vole sitting in a typical 'hunched' position whilst eating. This photo shows the blunt face, indistinct ears, small eyes and dark brown fur, with a slightly paler underside. Note also the standing leaves of cut vegetation around it. © Simon Booth Photography.

Figure 3.2   Black form of water vole being handled after being captured. © Stefanie Scott.

Figure 3.3   Water vole with its tail visible. © Jo Cartmell.

Figure 3.4   Water vole swimming. © Jo Cartmell.

4. Looking for field signs

Different field signs

There are a number of different field signs to be searched for when undertaking water vole surveys. Some of these are characteristic only of water voles, and therefore allow us to confirm presence of the species. Others are more easily confused with other species and therefore require some caution when interpreting them.

Water vole **droppings** fall into the first category – they are characteristic, and can be distinguished from those of other species. During the breeding season they are deposited at specific sites, called **latrines**. This is also characteristic of water voles. Droppings and latrines are therefore the best field sign for confirming the presence of water voles, and I'll discuss these at length in Chapter 5. Latrines can also provide useful information on the approximate relative density of water voles, and therefore give us clues as to how water voles are using different sections of a watercourse, for instance.

Feeding remains and **burrows** are also useful field signs, although these can be more easily confused with those of other species, particularly by the less experienced surveyor. I'll cover these in Chapters 6 and 7 respectively.

Water voles also construct above-ground **nests** in certain circumstances. However, they are not always present and they can be difficult to find and identify even when they are. They are therefore not as useful as a field sign, but it is important to note them where they are recorded. I'll discuss nests in Chapter 8.

Footprints and **runs** through the vegetation can also be useful field signs in some situations (see Chapter 9). However, it is difficult to distinguish water vole footprints from those of rats, and runs through the vegetation can be left by a range of different mammals. These signs (or their absence) are therefore only really useful when interpreted alongside other field signs. For example, a few years ago I was asked to provide advice on the mitigation required for water voles at a particular development site. The field signs of water voles recorded in that case were piles of very small droppings – so small that I believed that they were actually field vole rather than water vole (see Chapter 5). And the absence of obvious runs through the vegetation connecting the piles of droppings was useful evidence that convinced me – the bankside vegetation was so dense that, had water voles been present, there would have been runs present.

I often read survey reports that refer to another characteristic field sign of water voles – the 'plop' sound that they make when they dive into the water. Having spent the best part of 25 years surveying for water voles, I'm not at all convinced by this. It's true that when you approach a water vole that is sitting on a bank, just above the water, it is reasonably likely to dive in to get away from you. Or when you're wading along a stream, an animal sitting within the floating raft of vegetation that you're about to disturb will dive into the water and swim away. And they do make a noise when this happens. Is it characteristic? Can I tell from the sound alone that it's a water vole rather than something else? No, I can't. Someone else may be able to, but I'm afraid I'm sceptical about this.

If I heard a noise like this when doing a water vole survey I would stop, keep still and watch for an animal swimming away, or look for movement in the emergent vegetation where an animal has climbed out of the channel having already swum away. And then I'd approach the spot where I heard the sound come from and search for field signs there. That doesn't make the sound a field sign. I'm simply using my ears when I do a survey for water voles, as well as my eyes. I'll use my nose as well in some situations – but more about that later.

When to search

You can find field signs of water voles at any time of year. However, water voles are relatively inactive during the winter months (typically October/November to February/March, depending on where you are in the UK and what the prevailing weather conditions are like). This means that the best time to search for signs of water voles is the period when they are breeding, which in lowland habitats is effectively from late spring, through summer and into early autumn (generally mid-April to the end of September across much of the UK). There are parts of the UK where water voles have a longer breeding season (such as southeast England), and you can thus find plenty of evidence of them in March or October, but there are other locations where the breeding season is shorter, and you'll probably find very little except during May, June, July or August (such as in parts of Scotland or upland sites in northern England), or see an even shorter window at upland sites in Scotland.

May and June are generally good months for undertaking water vole surveys in the lowlands. As summer progresses, the vegetation will become more dense. This can make surveying more difficult in some situations, particularly in narrow, steep-sided ditches. However, where access is unrestricted to the places where field signs of water voles are likely to be, surveys later in summer and in early autumn (August and September) can be particularly good, as young animals born from the earliest litters of the season will be starting to set up territories, and there is therefore likely to be more activity, increasing the likelihood of finding field signs.

It's important to remember that water voles might use different parts of a habitat at different times of year. For example, a ditch that regularly dries out in summer might be used in spring but not in autumn, with the animals moving into other habitat nearby that still holds water. And the vegetation along some watercourses may not provide sufficient cover until mid-summer at the earliest, with the result that a late spring survey doesn't record them, whereas a late summer survey would. A water vole colony will also expand its range (and/or will increase in density) in late summer or autumn, because of the increasing numbers of animals present. It is for these reasons that the current guidelines for undertaking water vole surveys to assess the potential impacts of a development project on water voles (*The Water Vole Mitigation Handbook*: Dean *et al.* 2016) recommend two survey visits – one in the first 'half' of the season (mid-April to end of June) and one in the second 'half' (July to September inclusive).

If a survey undertaken outside the optimum season (i.e. between October and mid-April in much of the UK) fails to record signs of water voles, its results should be treated with caution. And whilst surveys at this time of year can confirm presence, they are unlikely to provide a reliable indication of relative population density or a full indication of distribution.

You should also try to avoid looking for signs of water voles during periods of high water levels or heavy rain, or in the week or so immediately following such events. This is because water vole latrines and feeding remains will become washed away in either case, and will take time to be re-established. Burrows may also not be visible during high water levels and it will be difficult to assess the suitability of the habitat (see Figure 2.10).

It is also worth remembering that surveys very early in the season (mid-April to early May in lowland England, for example) can sometimes fail to record field signs of water voles, particularly where they are present at low densities. I would be cautious about the interpretation of negative survey results in such circumstances.

Where to search

In my experience the vast majority of the field signs of water voles are found within 1–2m of the toe of the bank – that's the point where the bank meets the water (or vice versa). This gives you a linear search corridor in most cases, centred on the toe of the bank, extending 1–2m up the bank, and 1–2m out into the watercourse or waterbody.

Where there is a very wide fringe of marginal vegetation you'll have to look further from the bank edge. Water vole field signs could be found at any point within vegetation like this, so you might have to search all of it. And in some cases you might find signs well up the bank, further away from the water's edge, although water voles rarely use terrestrial habitat, particularly in lowland situations (there are exceptions, as already discussed). In my experience the likelihood of finding field signs drops significantly if you search more than 2m from the water. This is supported by Moorhouse *et al.* (2008), who found that 99% of water vole feeding remains, at a range of study sites, were located within the emergent vegetation (within the channel itself) or within the first 1m of bankside vegetation. In circumstances where there are field signs more than 2m from the water's edge, I would normally expect to find some closer to the toe of the bank as well.

There are also situations where water voles are present, but there's no evidence at the toe of the bank. Many years ago I confirmed the presence of water voles in two large ponds, relatively close to each other. Previous surveyors had failed to find them because, I believe, of where they looked. The ponds both had very shallow sloping banks, with little herbaceous vegetation on them due to the presence of willow scrub around their edges. There were no field signs of water voles around the edges of either pond. However, the ponds had a number of large tussocks of rushes in them and, in some places, extensive beds of reedmace. These were the places to look for signs of water voles, as these were the parts of the pond that the animals would be most likely to use. And, sure enough, the signs were there – it was just a case of working out where to look.

Bear in mind also that water voles use habitats away from water to disperse, sometimes crossing several kilometres of terrestrial habitat. However, finding evidence of them in such situations is likely to be challenging, if not impossible.

How to search

Firstly, you need to work out how to get the best view of the locations where water vole field signs are most likely to be. In some cases you can do this from the bank top. However, in the majority of cases, particularly where the vegetation is tall or overhangs the toe of the bank, you'll need to be in the water. Depending on the nature of the watercourse or waterbody you're surveying you might be able to do this wearing wellies, or you might need to wear waders, or in some cases you'll need to be in a boat.

Of course, if you're not trying to find every single field sign then you may be able to survey from the bank top, only getting into the water in places to examine specific features. In this case it is worth using a pair of binoculars to examine the toe of the opposite bank – but be aware that you won't find all of the signs using this approach, and in cases where the overhanging or emergent vegetation is dense you might not see them at all.

Secondly, you'll need to focus on the most likely places. This comes with experience, but it won't take long to get the hang of it. You'll be drawn towards tall emergent vegetation and you'll gently part it to look inside for evidence. Or you'll carefully lift up the grasses overhanging an undercut section of bank to get a view of the toe, where field signs are most likely to be.

Thirdly, you'll have to 'get your eye in'. Searching for small greeny-brown droppings against a backdrop of brown earth (or sometimes greeny-brown earth), or vegetation that's been cut and piled in a certain way against a backdrop of the exact same vegetation, is easy once you've got the right search image. However, I know from the experience of running training courses that someone without the right search image might not see those droppings or piles of cut vegetation, even when they're staring right at them from less than a metre away. But don't worry – this comes with experience, and hopefully the photos in this book will help.

Fourthly, you need to move methodically and carefully, searching all of those likely places. If you're wading in the water you'll need to make sure that you don't accidentally wash the field signs away before you get to them. This means moving steadily and carefully, particularly when approaching those locations to search. In a watercourse it's better to move upstream, as you're less likely to wash signs away, and it's also safer (see *Health, safety and biosecurity*, below). You should also make sure that you aren't damaging the habitat excessively, aren't squashing burrows, and aren't destroying nests.

Burrows can be very easy to damage if they are in shallow sloping banks, where the surveyor may tread on them, so extra care needs to be taken in such situations. Particular care also needs to be taken when surveying areas used by terrestrial, or fossorial, water voles; their burrows can be very shallow (typically between 10cm and 60cm deep) and are therefore easily damaged by walking over areas used by these animals.

Other factors to consider

When surveying for water voles it is important to have a good understanding of the context of the habitat you're searching. Are there known records of water voles from the general area? Is the habitat being surveyed typical of what's in the surrounding area, or relatively rare? And how well linked is the habitat being surveyed with other suitable habitat? It's easy to turn up and search a given stretch of watercourse for signs of water voles, but knowing how this stretch fits in with other habitat available in the wider landscape is critically important. A desk study, including a search for existing records, along with a review of Ordnance Survey maps and aerial photos to get a broad-brush understanding of likely habitat suitability and habitat links, is a sensible starting point.

You will also need to have a good understanding of how a watercourse or waterbody is managed to allow interpretation of the information – see Figures 2.20 and 2.21, which demonstrate this point. Who undertakes the management? How frequently are the banks mown? How often is the watercourse dredged? Are refuge areas left intact during management operations? This might all be obvious during the survey. But then again it might not – so making some enquiries in advance of doing the survey will be important. And it will be pretty fruitless turning up to do a water vole survey in the days or weeks immediately following management being done, as few field signs will remain, and they may take time to be re-established.

It's also vital that you know how water levels vary on a given watercourse. Again this might be obvious during the survey if you see flood debris piled up, but this won't always be the case. Local knowledge will be vitally important here. How often does the watercourse overtop its banks? Do certain ditches dry out in most years? How much do the levels typically vary?

Health, safety and biosecurity

It's hopefully fairly obvious that, when surveying wetland habitat, you'll need to consider your approach carefully, to minimise the risks involved. This will mean thinking about the clothes and equipment you wear and carry, where precisely you're going to walk, and how you're going to be rescued if the worst should happen. I'm not going to tell you everything you need to think about here – you'll need to consider that on a case-by-case basis yourself, even reviewing it whilst you're doing the survey. However, here are a few things to think about as a starting point:

- Is it safe to work alone, or do you need to be accompanied? Many companies and organisations undertaking water vole surveys will insist on there being two people present at all times.

- Do you need to wear a buoyancy aid or life jacket, and, if so, what sort?

- If a second person on the survey is there to help get you out if you have a problem, how will they do this? Do you have a rope or throw line?

- Are you able to summon help if needed? What's the phone signal like? Where would emergency services need to get to?

- Is it safe to be in the water? How fast is the current? Is there any obvious sign of pollution? How clear is the water (so that you can see what's below the surface)? Is there a weir downstream that could represent a hazard if you lose your footing and the current takes you towards it? Could there be hidden dangers, such as deep holes in the riverbed, submerged branches, or areas of deep silt?

- If you're getting into the water, how easy is it to get out again? If I wasn't confident about being able to get out of a watercourse, I wouldn't get in to start with.

- How deep is the water? Do you need to wear waders or wellies? Chest waders can be dangerous if you slip over, as they can fill with water and drag you under. There will be designs you can choose to avoid this. You might want to think about using a boat instead.

- What is the bed of the watercourse like? Is it silty, which will cause you to sink in and hamper your movement? Is it stony, and therefore potentially slippery?

- If you're walking along the bank, how steep and slippery is it? Is it undercut, and therefore prone to collapse if you stand on the very edge of it to look for field signs? Steep and slippery banks are best avoided – sometimes it's safer to be walking in the water, rather than walking on the bank with the risk that you might fall in.

- How are you going to minimise the likelihood of slipping over once in the channel, or the risk of getting out of your depth? I always take a trekking pole with me on water vole surveys. You can use it to test depth. They often have a helpful set of measurements on them, but you can also adjust the length of the pole so that a specific point on it is the depth you're comfortable getting to. You can also use a pole (or a strong stick) to test the stability of the bed, to help you balance in faster-flowing water or when standing on the bank, and to help you get safely in and out of the channel. And they're also good for parting vegetation that's a little way in front of you.

- Remember that water voles, and other animals associated with watercourses, can carry Weil's disease (leptospirosis), so you'll need to take appropriate precautions to minimise the risk of catching this disease and be familiar with its symptoms.

- Where are you going to start and finish your survey? In a flowing watercourse, where I'm planning on wading in the channel, I would generally start at the downstream end and work my way upstream. On stony riverbeds you are less likely to slip if you walk against the current. And on silty streams the sediment you stir up will be washed downstream – this means that it won't obscure your view of where you're about to take your next step, which it would if you were wading downstream.

It's also vital that you undertake appropriate biosecurity measures when surveying for water voles. This means taking steps to minimise the likelihood of you transferring plant material or diseases hazardous to wildlife such as crayfish plague, for example. Again, I'm not going to tell you everything you need to think about here, but making sure you clean and thoroughly dry all of your clothing, footwear and any other equipment before you survey a different watercourse or waterbody is essential. You should also make sure that you remove any obvious plant material from your clothing and equipment as soon as you exit the water. You can use appropriate disinfectants to clean clothing and equipment if you don't have the opportunity to dry them out.

5. Droppings and latrines

Water vole droppings are distinctively different from those of other species that you might find. Once you become familiar with the various characteristics to look for it will be relatively straightforward to confirm identification of the species (Figures 5.1 to 5.3). That is, of course, assuming you find some!

Characteristics of water vole droppings

Size	Approximately 9–15mm in length and 3–5mm in diameter.
Shape	Regular and even in shape, rounded in section, relatively broad, usually with blunt or rounded ends (but occasionally with one slightly pinched end).
Colour	Most are olive-green in colour and, when broken in half, are a lighter green on the inside; the colour can vary though, depending on what the animals has been eating – I've seen brown, purple and khaki droppings, although these are rare.
Consistency	Fairly solid; can be broken open easily to reveal chopped-up vegetation on the inside and, in the vast majority of cases, nothing else. Wet when fresh, drier as they age.
Smell	A faint earthy, compost type of smell.

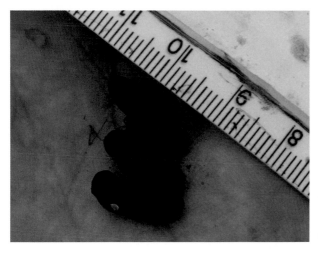

Figure 5.1 Relatively large water vole droppings (14–15mm long), which are very fresh. Note the regular, even shape, with rounded ends.

Figure 5.2 'Standard' size water vole droppings (10–12mm long), also relatively fresh. As in Figure 5.1, note the regular, even shape and the rounded ends.

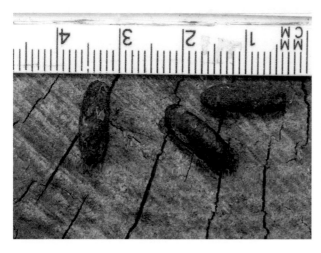

Figure 5.3 Water vole droppings that have dried out. The droppings become more solid when dry but maintain their regular shape and (generally) olive green colour. The length of time taken to dry out will vary, dependent on the prevailing weather.

Latrines

During the breeding season (see Chapter 4, section on *When to search*) water voles leave discrete piles of droppings as a territorial mark. These are deposited regularly throughout an animal's territory and are called latrines. The water vole will return to the same spot regularly, meaning that there are often droppings of a variety of ages present.

And, not content with leaving a pile of poo as a means of advertising its presence, the animal will then stamp scent onto the droppings using its hind feet, which it pushes past scent glands on its side. This isn't a scent that humans can detect, so isn't that useful as a means of identifying the presence of water voles (although dogs have been trained to identify it). Nevertheless, the act of the animal stamping its scent onto the droppings does cause many of the older droppings in a latrine to be clearly squashed, and eventually the earth in the area of a latrine that has been present for a while will turn olive-green in colour, as it is effectively just a mass of squished water vole poo (Figures 5.4 and 5.5).

Fresh droppings   Older dropping

Green 'mush' of squashed droppings, from the animals stamping on them – or, in this case, perhaps from recent heavy rain

Figure 5.4   Water vole latrine.

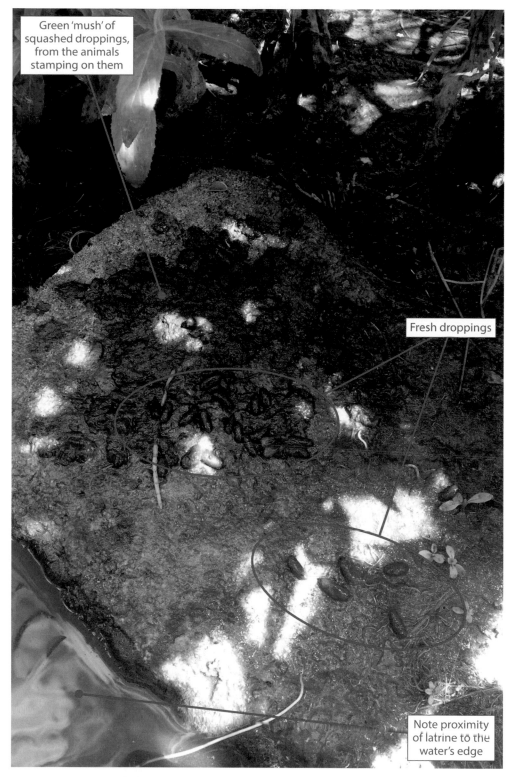

Figure 5.5 Water vole latrine on a flat rock.

Latrines, then, are a characteristic field sign of a water vole. Rats will leave droppings next to a watercourse that are of a similar size to those of water voles, and there can be many rat droppings left on the same rock, for example. However, rats don't create such extensive piles and don't stamp scent onto the droppings, and therefore don't create latrines.

In some cases latrines will be used by more than one animal. This can be difficult to detect unless there's a clear difference in the size of the droppings within a latrine, or distinct variation in the colour of droppings of the same age (Figures 5.5 and 5.8).

Water voles establish latrines in a variety of locations. They are normally in close proximity to the water, often within centimetres of it, or even in it. An obvious feature may be used, such as a rock (Figures 5.5 and 5.8), a tussock, a floating or handily placed piece of wood (Figure 5.9), floating polystyrene (Figure 5.10) or trampled vegetation. It may simply be a slightly raised section of bank (Figure 5.11). Sometimes they are on a ledge formed by flat stones used to reinforce a bank (such as in Figure 2.15), on top of a drainage pipe, or in the entrance of a burrow. An experienced water vole surveyor's eye will be drawn to the places most likely to be used as latrine sites – this comes with practice. It's incredibly useful to give you confidence in a negative survey result, when you don't find evidence of water voles, but you know you've seen multiple locations where you would expect to have found a latrine if water voles had been present.

Figure 5.6 Water vole latrine showing variation in colour of droppings. It is difficult to tell if the droppings are from different individuals in this case, as they may be of different ages.

Figure 5.7 Water vole latrine showing variation in size of the droppings, demonstrating that the latrine is used by more than one animal. The smallest droppings are approximately 9mm in length whereas the largest are 12–14mm long.

Figure 5.8 Water vole latrine on a piece of flat rock at the edge of a stream. Note the proximity of the droppings to the water.

Figure 5.9 Water vole latrine on an old timber fencepost that has fallen across a stream.

Figure 5.10 Water vole latrine on floating polystyrene.

Figure 5.11 Water vole latrine on a raised section of bank, outside a burrow.

How many droppings constitute a latrine?

This is a question I've been asked more times than I care to remember. If you find one dropping, is that a latrine? What about two? Five? And so it goes on. And what if there are only a few fresh droppings and no old droppings stamped into the ground?

My answer to these questions (perhaps unhelpfully) is that you can't define a latrine based on a minimum number of droppings, and that evidence that droppings have been left at the same location for weeks or months is not always visible.

The reason for this is that whilst some water vole latrines persist over time, others are much more ephemeral in nature. They may be so close to the water's edge that a small rise in the water level may wash the latrine away. The animal returns to leave some more droppings after the level has dropped, but only the fresh ones are visible. Or it creates a new latrine nearby – again, you may only see fresh droppings.

In some cases a latrine may be floating in a pond (Figure 5.12) or balanced precariously in a way that means it could be lost if disturbed (Figure 5.13). These may be present one day and gone the next. And a surveyor that isn't concentrating during the survey could easily cause them to be lost.

Figure 5.12   Water vole latrine on the surface of a pond with feeding remains.

Figure 5.13 Water vole droppings balanced on a piece of partially 'felled' and partially floating reedmace.

Figure 5.14 There are only a few water vole droppings visible in this photo, but there are some that are clearly older than others, and they are in a location where the vegetation has been worn away by regular activity (almost certainly due to frequent visits by water voles). I would record this as a latrine in these circumstances, irrespective of the total number of droppings present.

It's worth thinking about the location you find the droppings in. Is it somewhere that you would expect to find a water vole latrine? Has the vegetation been worn away around it, suggesting regular activity by an animal in that area? Has there been heavy rain recently that could have washed away older droppings, or caused a change in water level that has submerged previously used latrine sites? Does the location of the droppings itself make them unlikely to persist for long?

In many cases I'd argue that even a few droppings is a latrine. If they've been left by an animal to mark its territory, then it's a latrine (see Figure 5.14, for instance).

Why does this matter? Well, for the purpose of confirming the presence of water voles it doesn't, but we might infer something from the number of latrines recorded (see below).

What can latrines tell us?

Latrines are the most distinctive field sign of water voles, and are reliably present during the breeding season. We can therefore confirm the presence of water voles on the basis of finding latrines and no other field signs. The likely absence of water voles is a little more difficult – there are instances where you might find other water vole field signs but no latrines, such as where water voles exist at very low density, or where there is a lack of good latrining sites, or it is difficult to access the locations where latrines are likely to be present. And this, of course, assumes that you've done the survey at the appropriate time of year (see Chapter 4).

The number of latrines can also tell us something about relative population density. The more latrines you find in a given length of habitat, the higher the density of water voles using that habitat (see Dean *et al.* 2016). There is a published formula for calculating absolute numbers of animals from the numbers of latrines, but the estimates generated by the formula have very wide confidence limits; in addition, there are unknowns about its applicability to large populations and scenarios where water voles are using both banks of a wide watercourse (Morris *et al.* 1998).

One of the problems with using a formula like this is that there may be a difference in what any given surveyor considers a latrine. And other factors may affect latrine size and number, such as population density and the season when the survey was undertaken.

In general, the density of latrines allows a useful comparison between sites, or between parts of a site, or even between the same site in different years, provided that there is consistency in terms of the time of year when the survey was done, the level of effort invested in searching for latrines, and the interpretation of what constitutes a latrine. Calculations to estimate numbers of animals present based on the number of latrines recorded should be treated with caution.

Terrestrial, or fossorial, water voles

Water voles in terrestrial habitats will also deposit droppings in latrines that resemble those shown in the preceding figures. However, these latrines are often left on small mounds of earth which the animals create, superficially resembling mole hills but more compacted (Figures 5.15 and 5.16). Or, equally likely, they may be left under a grass tussock.

Figure 5.15 Terrestrial, or fossorial, water vole latrine left on a small 'mole hill' of earth. © Robyn Stewart.

Figure 5.16 Terrestrial, or fossorial, water vole droppings. © Robyn Stewart.

Rat droppings

Rat droppings are superficially similar to those of water voles, as they can be roughly the same size, and are often found deposited on obvious features on the banks of a watercourse.

One common difference between rat droppings and those of water voles is that water voles normally deposit their droppings in a latrine (during the breeding season). Rats don't do this. A rat may leave multiple droppings on the same 'feature' such as a rock or an exposed section of bank, but they tend to be deposited as individual droppings next to each other, rather than as a mass of droppings. Of course, in some cases you might only find a few droppings, which can make life a little more difficult.

Part of the problem with identifying rat droppings is that they are very variable because of the animals' omnivorous diet, making it difficult to generalise about their characteristics in the same way that we can for water vole droppings. It should be possible to distinguish them from water voles, though, by comparing across all of the characteristics listed below. Rat droppings normally differ from water vole droppings against at least one characteristic, and often more than one.

Characteristics of rat droppings in comparison with those of water vole

Size   Variable. Approximately 10–20mm in length and 3–7mm in diameter. Some are therefore clearly larger than water vole droppings and can be identified based on size. However, there is a significant amount of overlap in size and there will therefore be some that cannot.

Shape   Variable. Rat droppings are often irregular in shape, with tapered ends unlike those of water vole droppings (Figure 5.17). However, they can also be regular and even in shape and therefore superficially resemble those of water vole (Figure 5.18).

Colour   Variable. Rat droppings are normally black or brown. Water vole droppings can be similar in colour, and it may therefore be difficult to confirm that a given dropping is rat rather than water vole. However, rat droppings are rarely olive-green, and therefore olive-green droppings are far more likely to be from water voles.

Consistency   Tend to be less solid than water vole droppings. Difficult to break open, as they 'squish' when handled.

Smell   Repulsive! When fresh, a rat dropping has a very unpleasant smell. I'm not sure how else to describe it, but it is completely different from the earthy, compost type smell of a water vole dropping. If you smell it and recoil in disgust, then it's almost certainly a rat poo!

So, you will be able to identify rat droppings in many cases based on their large size, or their tapered, irregular shape. Colour is less helpful, given its variability in both water vole and rat droppings. And the variation in the size, shape and colour of rat droppings means that you are likely to find droppings that are not easy to

Figure 5.17   Rat droppings, deposited on bank protection next to a stream.

assign as either water vole or rat based on these characteristics alone. The squishy consistency (I might also describe it as 'a bit icky') and the disgusting smell are very helpful characteristics. You should be able to distinguish rat droppings from water voles on the basis of a combination of these factors.

Figure 5.18   Rat droppings, deposited on a reinforced section of bank next to a stream. The photo shows two fresh droppings next to a 20p coin for scale; to the left is a fragment of a possibly older rat dropping.

Field vole and bank vole droppings

The droppings of field voles and banks voles are also quite easy to confuse with those of water voles. This is because the droppings of these smaller vole species are just a miniature version of the dropping of a water vole. So there's not really any point in discussing shape, colour, consistency or smell for field vole and bank vole droppings, as these are no different from what I've already described for water voles.

Both field and bank voles, but particularly field voles, will occupy similar habitat to water voles (as well as habitat that water voles don't normally use). They will therefore leave droppings in similar locations to water voles. Furthermore, field vole droppings are frequently olive-green in colour and left in small piles which resemble a water vole's latrine (Figure 5.19), and are often in among feeding remains (Figure 5.20).

The only obvious difference between water vole droppings and those of these smaller vole species is size. Field vole and bank vole droppings are shorter (5–8mm) than a water vole dropping (9–15mm) and are also narrower (2–3mm) than those of water voles (3–5mm). Although these measurements suggest that the droppings of smaller vole species may be only a few millimetres shorter and narrower, the combination of being both shorter and narrower means that the overall size (volume) of these droppings is noticeably less (Figure 5.21). Again, it's about getting your eye in to notice this difference.

Personally, I tend to find it very difficult to pick up and manipulate the droppings of these smaller vole species between my fingers. They're too small and get squashed in the process.

I've often read survey reports that refer to the discovery of latrines deposited by 'baby' water voles. Now, it's true to say that young water voles, such as those animals born from the first litters of a season and starting to set up territories by the end of the season, will leave relatively small droppings. However, you're unlikely to find a colony consisting just of young animals. I strongly suspect that the 'baby' water voles referred to by those survey reports were actually field voles.

It's always worth looking at the area where you find a pile of small droppings that look superficially like a water vole latrine. If an animal of the size of a water vole has been visiting that spot regularly it will have worn away the vegetation to a certain extent, and there are likely to be obvious runs along the bank linking the latrine sites (see Chapter 9). If this level of activity isn't visible, it's more likely that the droppings have been left by a much smaller animal – a field vole in most cases.

Figure 5.19   Field vole droppings left on a ditch bank and resembling a water vole latrine.

Figure 5.20   Field vole droppings found among feeding remains. © Magnus Johnson.

Figure 5.21 Water vole and field vole droppings in the hand, for size comparison.

DNA analysis

You might still find yourself in a situation where you are uncertain about the identification of a specific dropping. Perhaps you have very little other evidence to help you piece the puzzle together, and you're left with very little to go on. It is now possible to confirm identification through DNA analysis of droppings. So, if you're unsure, it might be worth collecting the droppings and having them analysed. Of course, you might go on to find so much evidence on the remainder of the survey that you don't need to worry about the identification of one specific dropping – it might make no material difference to the outcome. In which case DNA analysis is unlikely to be worthwhile.

It's also worth mentioning here that, at the time of writing this book, research is ongoing into the possible use of environmental DNA (eDNA) for confirming the presence of water voles in a watercourse or waterbody (see Halford *et al.* 2018, Halford and Moreton-Jones 2018). eDNA is genetic material isolated from the environment, which in the case of water voles means a water sample. There are, however, some limitations to the technique, and further research is required to demonstrate its efficacy.

6. Feeding remains

Water voles often leave discrete piles of cut vegetation, referred to as 'feeding remains' or 'feeding stations' (Figures 6.1, 6.2 and 6.3). These piles are normally left close to the water's edge and within areas with plenty of cover. Water voles will take cut vegetation to a preferred feeding location to eat it, although I doubt that this is what's happening in the vast majority of cases. A more common scenario, in my experience, is that the animals are simply sitting in among the vegetation that they want to eat, whether bankside grasses or emergent reeds or reed sweet-grass for example, and eating whatever is within easy reach. The places to look for water vole feeding remains, then, are in among the vegetation, close to the water's edge, along runs, and generally in locations where you might also expect to find latrines (see Chapter 5). Latrines and feeding remains are not routinely found in the same specific location, although they can be within 10–20 cm of each other.

Water voles will eat pretty much any bankside or emergent plants. In some cases the piles appear to comprise sections of a plant that an animal has deliberately cut into lengths, in preparation for eating them, presumably where it has been disturbed before it had a chance to do so. You will find sections of the stalk of reed sweet-grass or common reed, rush or sedge, or countless other plants. In other cases the piles of vegetation comprise the less palatable parts of a stalk or leaf that the water vole didn't want to eat, such as the long and relatively dry leaves of reedmace (Figure 6.3), or the tops of nettle or willowherb.

Figure 6.1   Typical water vole feeding remains at the water's edge.

Figure 6.2   A flattened area of vegetation where a water vole has been sitting to eat.

Figure 6.3 Typical water vole feeding remains (of reedmace).

As far as we're concerned when trying to identify the presence of water voles as part of a survey, the reason why an animal has left a pile of vegetation behind is relatively unimportant. What is important is to identify the species of animal. How can we tell that it was a water vole that left it, rather than something else?

There are some characteristics of the vegetation that we can look at, but we have to treat them all with an element of caution. This is because smaller voles, such as field voles, will also leave behind piles of chopped-up vegetation in similar locations, and this can cause confusion. Moreover, water vole feeding remains can vary markedly dependent on what the animal has been eating – as shown by the varied nature of the remains shown in Figures 6.1 to 6.13.

There are a number of things we need to do when we find feeding remains in order to allow us to identify who has left them behind.

Step 1: Base your assessment on a number of different pieces of vegetation rather than a single piece

Even within a single pile of vegetation there will be variation. It's always worth picking up a selection of pieces to examine more closely. Figure 6.4 shows an example of such a selection.

Step 2: Look at the lengths of vegetation

A number of guides on water vole field signs will refer to lengths of vegetation being up to 10 cm long. In reality you will find much longer lengths than this (30 cm or more), and you will also find much shorter lengths (1–2 cm). If all the vegetation lengths (or the vast majority) in a pile are:

- 30 cm or more in length – it's pretty likely to be water vole (Figures 6.5 and 6.6).
- 10–30 cm in length (such as in Figure 6.7) – that is a useful indicator, but it isn't sufficient on its own to confirm water vole, as those smaller voles have been shown to leave behind vegetation of similar lengths (Ryland and Kemp 2009).
- Less than 10 cm in length – you still can't rule water voles out I'm afraid (Figure 6.8).

Step 3: Look closely at the cut ends

Water voles will often leave the ends cut at a 45-degree angle, with the imprint of two large front teeth, and a small tear of vegetation where the last bit has been ripped rather than cut (Figures 6.4, 6.9 and 6.10). If you see this pattern, as shown by the top of the piece of vegetation on the right-hand side of Figure 6.4, then you can be reasonably confident that it's from a water vole.

But then again, they don't always do this. You will find vegetation that has been cut straight across, and not all cut vegetation will show that tear (again, see Figure 6.4).

Those smaller voles will also cut vegetation at a 45-degree angle. And, whilst the tooth marks will be smaller, they aren't always visible.

You also need to be careful about identifying water voles by looking at the cut ends of the vegetation that is still growing on the bank or within the channel of a watercourse, if there is no pile of feeding remains present. When water voles cut the stems they will often leave the same 45-degree angle as on the piles of 'felled' material. However, as with the piles of feeding remains, you won't always see this. And remember that other animals will eat vegetation in similar locations, such as wildfowl, and this can leave a similar pattern on the cut end of the material that is still growing.

Step 4: Consider the size of what the animal has eaten

Water voles are likely to take on a much larger piece of vegetation than a field vole (or a bank vole). So, it's always worth thinking about what the animal has eaten based on what it's left behind. Again, a water vole will eat small grasses and rushes just as a smaller vole species will. However, those smaller vole species are unlikely to be capable of 'felling' a tall stem of reedmace, reed sweet-grass, bur-reed or common reed, for example. They are more likely to pull down and eat the leaves where they can access them (although water voles will do this as well, particularly with reed sweet-grass).

Step 5: Search within the pile for droppings

Field voles will often leave droppings in among their feeding remains (see Figure 5.20). Water voles, in my experience, don't do this very often. So it's always worth having a careful search through any cut vegetation for droppings that might be hidden among it.

Step 6: Keep looking

If you still can't work out the answer then look for other piles of vegetation or other field signs.

Figure 6.4 Four pieces of vegetation taken from a single pile of water vole feeding remains. The piece on the right-hand side is 'textbook' water vole. The one on the left is not, and is only known to be from a water vole by association with the other pieces.

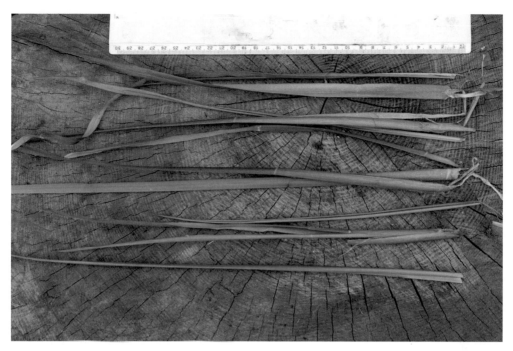

Figure 6.5 Typical water vole feeding remains in a bed of reed sweet-grass, with a 30cm ruler for comparison.

Figure 6.6 Water vole feeding remains (of reed sweet-grass) with a 30cm ruler for comparison. Not all lengths will be as long as this, but where they are it is reasonable to conclude that they are from water vole rather than a smaller vole species.

Figure 6.7 Water vole feeding remains (of common reed) with a 30cm ruler for comparison. Note the variation in lengths between approximately 8cm and 15–18cm. Lengths of this size are more likely to be from water vole than from a smaller vole species.

Figure 6.8 Feeding remains scattered at the water's edge. Note the variation in lengths, with some as small as 2cm. Some lengths also appear to show smaller tooth marks. Based on the location where they were recorded (where water vole presence had already been confirmed), and the presence of some longer lengths, these are all likely to be from water vole. However, the shorter lengths on their own could not be distinguished from the feeding remains of smaller vole species. It is important to examine multiple pieces.

Figure 6.9 Cut stem showing 45-degree angle, but not a clean cut, and the tear.

Figue 6.10 Close-up of the cut ends of water vole feeding remains, showing the 45-degree angle of the cut and the 'tear', but note that some teeth marks are visible when viewed this close.

Figure 6.11   Relatively old water vole feeding remains, where some of the cut vegetation has dried out in situ.

Figure 6.12   Water vole feeding remains.

Figure 6.13 Feeding remains from a terrestrial, or fossorial, water vole outside its burrow. © Robyn Stewart.

Figure 6.14 Field vole feeding remains found on a riverbank in a location where you would expect to find water vole feeding remains.

7. Burrows

Water voles excavate burrows in the banks of watercourses or waterbodies. Some of these are short 'bolt holes', extending no more than 30 cm into the bank. Others link to nest chambers and form a network of tunnels, potentially extending several metres from the entrance. Most water vole burrows have entrances that allow easy access to and from the water. Entrances may be fully or partly submerged below water level, or they may be further up the bank. Some burrows have entrances in the top of the bank, essentially linking a nest chamber within a section of bank 'vertically' to the habitat on the bank top. This is most likely to occur on watercourses or waterbodies with steep bank profiles and banks which 'top out' no more than about 2 m above normal water level. Much taller banks are less likely to have burrows extending vertically upwards.

Terrestrial, or fossorial, water voles also excavate burrow systems. These are likely to be very similar to those described above for water voles associated with wetland habitat, with the obvious exception that they don't relate to water. The burrows can be very shallow (typically between 10 cm and 60 cm deep). They can also be very extensive; one system, mapped using ground-penetrating radar, was found to comprise a series of tunnels radiating out from a large central chamber, with a total tunnel length of almost 40 m (R. Stewart, personal communication).

Some water vole burrows will be really obvious; they may be in a bare section of bank, or a bank may be so 'riddled' with them that it's starting to crumble in places (Figure 7.1). However, many are likely to be much less obvious; they may be hidden in dense vegetation, or have entrances below water level. You will have to lift up the grasses overhanging an undercut bank (Figure 7.2). You will have to carefully push aside tall emergent vegetation. And you'll often need to be looking at the face of the bank from within the water, as well as standing on top of the bank looking down onto it – difficult to do at the same time!

Burrows will often be easier to spot early in the survey season (April or May, for instance, in the lowlands), before the vegetation starts to hide them. But they will also become more obvious as they are regularly used throughout the survey season.

Figure 7.1 Water vole burrows – multiple entrances provide access to the same burrow system and may allow the animals to respond to changes in water level.

Latrine site

Evidence of activity on the bank and around the entrance

Figure 7.2 Water vole burrow at water level. The burrow was not visible until overhanging vegetation was carefully lifted out of the way, exposing the burrow and a latrine site next to it.

I read a water vole survey report recently that recorded sections of ditch (often several hundred metres in length) where no burrows were recorded but other field signs, including latrines and feeding remains, confirmed the presence of water voles. Now, if water voles are present in a section of ditch of this length, the likelihood is that there will be burrows (unless the animals are all using nests – see Chapter 8). This serves to underline the difficulty of finding burrows; it can be very difficult and time-consuming. Incidentally, I personally wouldn't distinguish in a report between watercourses where I've confirmed the presence of water voles and found burrows, and other watercourses where I've also confirmed the presence of water voles but not found burrows, unless I could be certain that burrows were absent – and this is very rarely the case.

Due to the difficulty of finding burrows, then, you can't rule out the presence of water voles if you don't record burrows.

So the next question is, can you confirm the presence of water voles if you do find a burrow? Well, yes, but only if you can be certain that it's occupied (or has recently been occupied) by a water voles as opposed to something else. And that isn't as easy as it sounds. Many different species create, or use, burrows of a similar size to a water vole burrow, and are associated with wetland habitat. We therefore need to look carefully at the characteristics of water vole burrows.

Characteristics of 'active' water vole burrows

Size	6–8 cm in diameter. They can get larger than this, particularly where there has been some erosion around the entrance.
Shape	Round. That sounds pretty obvious, but the noticeably rounded body shape of a water vole does mean that their burrows are also noticeably round in shape (Figures 7.3 and 7.4).
Level of activity	Active water vole burrows will have clear evidence of animals going in and out regularly. This might be in the form of 'puddled' clay around the base of the entrance, sides of the entrance that are worn smooth, or the wearing away of vegetation immediately around the entrance (Figures 7.5, 7.6 and 7.7). Sometimes a distinct 'swirl' of vegetation can be seen around the entrance (Figure 7.8).
Grazed 'lawns'	Water voles sometimes create grazed 'lawns' around their burrow entrances (Figures 7.10 to 7.12). This is formed from the animal poking only its head and shoulders out of the burrow and grazing what it can reach, even if that is just small grasses. It's not really known why they do this, but it may be done by females with dependent young below ground who don't want to leave their burrow system. If you see a grazed lawn, it's highly likely to be a water vole burrow (assuming that it meets all of the other characteristics listed above). However, not all water vole burrows have grazed lawns, so an absence of one doesn't tell you it isn't a water vole burrow.

Burrow plugs   Water voles will sometimes plug their burrows with what appears to be bedding material from the nest chamber. This is seen with terrestrial, or fossorial, animals, and less frequently with water voles in wetland habitat.

It's important to remember that burrow entrances that aren't in regular use at the time you find them might not show any of these characteristics other than 'size'. That doesn't necessarily mean that they aren't used by water voles:

1   If you find burrows outside of the breeding season then you wouldn't expect them to be regularly used, and they therefore won't necessarily show these characteristics; and

2   You still need to be cautious with a survey during the breeding season, as any burrow entrance you find may still form part of a burrow network, even if that particular entrance is not being regularly used and, in any case, not all water vole burrows display all of the characteristics.

For these reasons I think you can use the characteristics above to enable you to confirm that a burrow *is* likely to be a water vole burrow, but you can't use an absence of these characteristics (with the possible exception of 'size') to confirm that a burrow *is not* a water vole burrow.

Figure 7.3   Water vole burrow at water level, showing the characteristic roundness.

Figure 7.4 Water vole burrow just above current water level, at the time the photo was taken, but submerged in winter. Note also the characteristic roundness of the burrow.

Figure 7.5 Water vole burrow showing 'puddling' around the entrance.

Figure 7.6 Water vole burrows with worn-away area and a latrine outside the entrances.

Figure 7.7 Water vole burrow with clear evidence of activity immediately outside. The bankside vegetation has been worn away and a latrine and feeding remains are visible.

Figure 7.8 Burrow of a terrestrial, or fossorial, water vole. Note the round shape and the vegetation 'swirled' around the outside. © Robyn Stewart.

Figure 7.9 Water vole burrow, extending almost vertically up onto the top of the bank of a watercourse. There is some evidence of vegetation around the edges of the entrance having been nibbled.

Figure 7.10 Vertical water vole burrow, with grazed 'lawn', shown by bare earth patches and vegetation die-back.

Figure 7.11 Water vole burrows with grazed 'lawns' outside the entrances. © Gareth Harris.

Figure 7.12 Burrows of terrestrial, or fossorial, water voles, showing grazed 'lawns'. © Robyn Stewart.

Figure 7.13 Rat burrow. In excess of 10cm in diameter and with spoil outside the entrance.

Burrows of similar species

Of the other species that excavate, or occupy, burrows in similar habitats to water voles, the one most likely to be confused with water vole is brown rat (Figure 7.13). Rat burrows tend to be slightly larger, closer to 8–10 cm in diameter, although there is some overlap in size. They are also generally round in shape, but are less noticeably so. In general, they are more likely to be found in open, exposed locations, sometimes linked to other rat burrows by well-worn pathways. And they may also have an obvious soil heap outside the entrance. However, these are all generalisations, and where both species are present it can be very difficult to distinguish a water vole burrow from a rat burrow. That is, unless you find burrows with grazed 'lawns' – these are highly likely to be occupied by a water vole as opposed to anything else.

Field vole, bank vole and wood mouse (*Apodemus sylvaticus*) also create burrows in similar situations to water voles, but these are normally significantly smaller (up to 4 cm in diameter), and can be ruled out on size.

Rabbit (*Oryctolagus cuniculus*) burrows are generally taller than they are wide, and often have a spoil heap outside the entrance, sometimes with droppings on it.

Kingfishers (*Alcedo atthis*) use burrows that are normally higher above water level, in vertical bank faces, which the birds can drop out of.

Some species of crayfish will create burrows in the lower sections of the bank of a watercourse, around water level. These can be a similar size to water vole burrows, but there are often numerous burrows together in a short section of bank, normally in locations with little, if any, bankside or emergent vegetation. They are also generally not round in shape.

Detailed examination of burrows

It's generally not necessary to carry out detailed examination of burrows to confirm the presence or likely absence of water voles, as other field signs are normally found. It also isn't practical in many cases.

Internal examination of a burrow using a fibrescope, for example, is unlikely to provide much useful information. It may tell you that a burrow is a short bolt hole, rather than an extensive burrow system, but there are likely to be other burrows present that haven't been found, and such information is therefore likely to be of limited use. Personally, I don't carry a fibrescope with me on a survey as I can't foresee a situation where I'd be likely to use it (and use of it may require a licence – see below).

I'm aware that some people have set up remote cameras focused on a burrow when they are unsure if it is used by a water vole or a rat. This may provide useful information in circumstances where knowledge of which species is using one particular burrow is relevant. However, in general this is unlikely to be helpful as there will be multiple burrows, and proving that one is used by rats will not tell you which species is using the others.

It should be remembered that water voles are protected from intentional or reckless disturbance, and detailed examination of burrows (such as through use of a fibrescope or by setting up a remote camera nearby) may therefore be considered to constitute an offence under the legislation (see Chapter 1). You should determine whether a licence is needed before undertaking such examinations.

8. Nests

Water voles use nesting material below ground, within nesting chambers in their tunnel system. In certain situations they also construct and use nests above ground.

Above-ground nests seem to be used most frequently where there is a dense stand of emergent vegetation that is sufficiently robust to support a nest, such as reed sweet-grass, common reed or a large species of rush (as in Figure 8.1). They are also often found in situations where there is a lack of burrowing opportunities close to the water's edge; the animals construct the nest to allow them to exploit a particular part of the habitat, sometimes constructing it on top of floating rafts of vegetation (as in Figure 8.2). I've also recorded nests within large tussocks of rushes dotted through the centre of a shallow pond; the animals had eaten the vegetation in the centre of the tussocks, created a nest, and left the fringe of vegetation around the edges of the tussock intact, presumably as a screen to hide the nest and themselves.

I find nests at some sites in late summer or autumn, where there were none present in late spring or early summer. This might be due to the vegetation taking time to establish to a point where it is capable of supporting a nest. It might also be that later in the season, as water levels drop and emergent vegetation grows, the water's edge creeps further away from the bank which the animals had burrows in. At a certain point the animals may give up on the burrows, choosing to create a nest to reduce the distance they have to travel to access the vegetation directly adjacent to the water's edge. This is the case in the wide bed of reed sweet-grass along the edge of a river shown in Figure 2.6.

The nests are about the size of a small football, about 20 cm in diameter, and can be either round or more rugby-ball shaped. They are formed from the leaves of surrounding vegetation and do not appear to be particularly expertly woven. On the inside, the vegetation used to construct the nest is finer and appears to be pushed rather than necessarily woven into shape. For this reason it would normally be impossible to remove a nest from the vegetation without damaging it – something you shouldn't do in any case, as this may well constitute an offence under the legislation protecting water voles (see Chapter 1).

Water vole nests are difficult to find, particularly as you should be trying to minimise damage to the vegetation during a survey, and this will make it hard to find nests in the densest areas. And they are not always present. This means that you can't interpret anything from an absence of water vole nests. Also, it's reasonably unlikely that nests would be present in the absence of any other field signs of water voles, and you're therefore unlikely to be trying to confirm presence on the basis of a nest only. Nevertheless, it is important to record nests if you do find them, as this provides useful information about how water voles are using the habitat.

A number of different bird species will obviously construct nests in similar locations to water voles, although these tend to have an obvious hollow in the centre of them which a water vole's nest lacks. The nests of harvest mice (*Micromys minutus*), which you might also find in reedbeds, are significantly smaller and normally woven further up the stem of the vegetation.

Figure 8.1 Water vole nest among rushes. Nest constructed from fine material supported by the vegetation.

Water vole droppings

Water vole nest constructed from fine material on top of flattened reeds

Figure 8.2 Water vole nest among common reed.

9. Other field signs

Footprints

Water vole footprints have five toes on the hind foot and four toes on the fore foot.

The four toes on the fore foot are spread out in something like a star shape, with the outermost toes pointing at almost 180 degrees to each other. The fore print is, on average, 18mm long and 23mm wide (Couzens *et al.* 2017).

The five toes on the hind foot all point generally forwards. The three central toes are almost parallel with each other, with the outermost toes pointing out at an acute angle. The length of the hind print is given as 30mm long by Couzens *et al.* (2017), and as 26–34mm by Strachan *et al.* (2011). There is also a longer heel visible on the hind foot than on the fore foot.

The footprint of a rat is almost indistinguishable from that of a water vole (Figure 9.1). The footprints will obviously be larger in general, due to the size difference of the animals. Rats can have significantly longer hind prints (given as up to 45mm in Strachan *et al.* 2011). However, there is relatively little size difference on average (a matter of a few millimetres), and the exact size of a footprint can be misleading as it depends, to an extent, on the nature of the substrate within which it is registered – a footprint will often appear slightly larger in a softer substrate. This makes it difficult to be certain about identification from footprints alone.

Rat footprints are generally reported to have a longer heel registered in the prints from the hind foot, and there are reported to be discernible differences in the angle between the outermost toe and the next nearest toe. However, it is incredibly difficult to pick out these characteristics in the field and I've therefore not described them in detail here. This is because you rarely find a single line of well-registered water vole or rat footprints, as you might for an otter or a badger for example. Instead what you encounter is an area of soft ground that an animal has passed over many times, over-printing old footprints with new ones (Figure 9.2), and this can make it impossible to pick a single good footprint out of the mess!

As a general rule then, footprints are not a great field sign – they just tell you that you've got water voles or rats, or both.

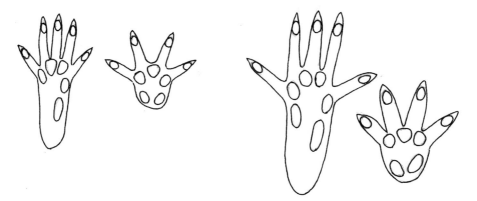

Figure 9.1 Water vole and rat footprints. From left to right: water vole hind foot, water vole fore foot, rat hind foot, rat fore foot (not actual size). Reproduced from *Water Voles* by Rob Strachan, published by Whittet Books, 1997.

Figure 9.2 Water vole footprints on an area of exposed mud; droppings are also visible. © Kevin O'Hara.

Runs through the vegetation

Water voles are a reasonable-sized mammal and do leave a clear run through the vegetation on the banks of a watercourse, for example, of about 6–10cm in width. However, many other animals will also create runs through the vegetation. So on its own this isn't a diagnostic field sign.

It is, however, useful in combination with other signs. If you find a run of the right size, then it's worth following it. If it's been made by a water vole, then it will normally link burrows, feeding remains and latrines (assuming that you're surveying at the right time of year – see Chapter 4), so follow the run until you find one or more of these.

The absence of runs through the vegetation can also be a useful thing to note. If a water vole was moving around on the bank, through dense grasses for example, then you'd expect to see a run somewhere (assuming again that you're surveying during the breeding season and when the vegetation is sufficiently well developed). I wouldn't rule out the presence of water voles based purely on not finding a run, but it can certainly help 'build the case' for likely absence, in combination with an absence of other signs.

And this can prove particularly useful where you've got field voles present on the banks of a watercourse, leaving small piles of droppings that look like miniature latrines. If the vegetation on the bank is dense, and there's no evidence of activity of an animal the size of a water vole moving through it, this will help you to be certain that those droppings are field vole and not water vole. But of course the two species do coexist in some cases, so be careful!

Rats will create well-worn pathways across open ground with no vegetation as cover. Water voles are less likely to do this. That's not to say that they won't in certain situations, but it's more likely that a pathway such as this has been made by a rat rather than by a water vole.

10. Other species

There's a reasonable likelihood that, during a water vole survey, you will encounter field signs of other species, or perhaps even see the animals themselves. Some of these species may well be relevant to the conservation of water voles – specifically brown rat, American mink (*Neovison vison*) and otter (*Lutra lutra*).

Brown rats may predate water voles, spread disease, and compete for space and resources with water voles. They also occupy burrows that look similar to those occupied by water voles, in similar locations. For these reasons it will always be worth noting evidence of the presence of brown rat during a survey. I've already discussed their field signs extensively, so won't cover them again here.

American mink is a non-native species that has been breeding in the wild in the UK since the late 1950s (Harris and Yalden 2008) and is a predator of water voles. They are a particular threat to water vole conservation, and it is important that water vole surveys record any evidence of their presence. Mink are well adapted to predate water voles (and water voles are, conversely, not well adapted to escape them). This is because mink can catch water voles in the water, as they are good swimmers, and a female mink is small enough to get into a water vole burrow. Water voles have no effective defence. Mink are also able to switch to other prey items once all the water voles are eaten, so there isn't a subsequent drop-off in mink numbers that would allow water vole populations to recover.

Otters may also prey on water voles, although they are likely to be less effective in doing so than mink. You may also confuse otter field signs with those of mink, and they are generally found in similar locations, so it's worth covering both here.

Droppings

Otter spraint and mink scats (as their droppings are respectively known) are often found on or near prominent features, such as under bridges, on rocks next to a weir or large rocks in a watercourse, or at confluences.

Otter spraint is black and sticky when fresh, drying out and then turning grey as it ages (Figures 10.1 and 10.2). When intact, an individual spraint is around 10 cm long and cylindrical (1–2 cm in diameter), but it breaks into smaller pieces easily. Spraint has a distinctive sweet and slightly fishy smell, which Rob Strachan likens to the smell of jasmine tea in *Mammal Detective* (Strachan 2010). It normally contains just fish bones and scales, and is therefore spiky when handled. However, otters will also regularly eat crayfish, crabs (coastal otters) and amphibians, and the appearance of the spraint will vary accordingly.

Mink scats are long, narrow and tapered with a twisted appearance (Figures 10.3, 10.4 and 10.5). They are black, dark green or brown in colour and foul-smelling when fresh. Mink have a varied diet, including mammals, birds and fish, and the scats will therefore contain fur and feathers as well as fish bones and scales.

Figure 10.1 Otter spraint deposited on a stone step next to a river. It contains fish bones and scales, as well as crayfish remains.

Figure 10.2 Very fresh otter spraint deposited on exposed tree roots, containing fish bones and scales.

Note the twisted, tapered end of the scat

Figure 10.3 Dried-out mink scat containing feathers and fur, with an old £1 coin (23mm diameter) for scale. © Gareth Harris.

Figure 10.4   Fresh mink scat deposited on a water vole latrine. © Coral Edgcumbe.

Figure 10.5   Fresh mink scat deposited outside a partially excavated water vole burrow. © Coral Edgcumbe.

Footprints

Both otters and mink register five toes in their footprints. The toes in both species are arranged around the pad, pointing in different directions (unlike the toes of badger (*Meles meles*), fox and dog, which all point forwards).

Otter footprints are larger than those of mink, being typically around 6 cm wide and 6–8 cm long (Figures 10.6 and 10.7). Mink footprints are generally around 4 cm wide and 4 cm long, although there is some variation (Figure 10.8). The toes in both species appear to taper to a point, as the claw normally registers as part of the toe, rather than separately. The toes of mink are noticeably narrower than those of otter.

Figure 10.6 A line of otter footprints in soft mud.

Figure 10.7 Two otter footprints, one partially over-printed on the other, with a 1p coin for scale.

Figure 10.8 Mink footprints on a clay pad taken from a mink raft, with a 20 pence coin (21 mm diameter) for scale. © Gareth Harris.

11. Recording the information

What information needs to be recorded?

Exactly what needs to be noted during any given water vole survey will depend on the purposes of that survey. What is the information going to be used for?

If you're undertaking a survey to inform an assessment of impacts, design of mitigation and consideration of licensing options for a proposed development project, then you should refer to *The Water Vole Mitigation Handbook* (Dean *et al.* 2016).

If you're undertaking a survey for other reasons then there is likely to be a specific protocol written for that specific survey, which will tell you what you should be recording, such as that described in the *Water Vole Conservation Handbook* (Strachan *et al.* 2011).

In general terms, I'd say that in all cases you should be recording the following, as a minimum:

1   Basic information about the survey
   - Surveyor's name.
   - Date of survey.
   - Survey location, including name of site/watercourse and Ordnance Survey grid reference(s) of survey extent.
   - Weather conditions during survey and within the week prior to the survey (in general terms).
   - Method of survey. Was the surveyor walking in the water, on the bank, using a boat, etc? Were both banks (of a watercourse) examined?
   - Any limitations to the survey. Was access restricted? Had watercourse management been undertaken recently prior to the survey? Could the weather have affected the outcome? Was it difficult to access the toe of the bank where water vole field signs were most likely to be found?

2   An assessment of the suitability of the habitat (for example, using Table 2.1 in Chapter 2) and, importantly, the reasons for your assessment. This is likely to include consideration of:

- Type of watercourse/waterbody (except for terrestrial, or fossorial, water voles).
- Bank substrate – suitability for burrowing.
- Likely variation in water levels relative to bank.
- Likelihood of drying out.
- Height of top of bank above water level.
- Approximate gradient of bank.
- Extent of bankside cover from herbaceous vegetation (and level of shading).
- Extent of marginal/emergent vegetation and, ideally, identification of the dominant plant species present.
- Extent and frequency of management, if known.

Note that, in some cases, it may be appropriate to divide a watercourse or waterbody into sections for the purposes of recording this information, particularly where there is variation within the overall length of habitat being assessed.

3   Latrines

- Recorded or not?
- Total number recorded, or approximate density (1 per 1 m of bank, 1 per 5 m of bank, etc.).
- Distribution within the length surveyed – are they evenly distributed, are they only in discrete areas, is there variation in the density, and how does this relate to the variation in habitat?
- Could any limitations to the survey have affected the number of latrines recorded, or locations where they were found?

4   Other water vole field signs

- Recorded or not? List any found.
- The total number of water vole field signs other than latrines is of limited relevance in most cases.
- The overall distribution of other water vole field signs is worth noting if they are found in areas where latrines weren't, but not otherwise.
- The specific locations of burrows and nests may be relevant and are worth noting, but be careful not to imply their absence just because you didn't find them.

5   Field signs of other relevant species

- Evidence of American mink, otter and brown rat is worth noting.
- Other species may be of interest but are likely to be of lesser relevance to the outcomes of a water vole survey.

How to record the information

In the field you should record on a suitably scaled map annotated with notes. You can use a standardised survey form (there's an example in the *Water Vole Conservation Handbook*). This will be particularly useful where you're trying to standardise the information being collected at multiple sites and/or by multiple surveyors. It won't be necessary in all cases.

How the information is presented in any final report will depend entirely on the aims of the survey, so I'll refer you back to *The Water Vole Mitigation Handbook* and the *Water Vole Conservation Handbook*. The one thing I'll say about this here is that it is vital that any limitations to the surveys are described and their implications explained.

Importantly, I would encourage you to take photos. Photos of the habitat will be incredibly useful – perhaps a 'typical' photo for each site, or each section you divide a survey area into. It's often also useful to take photos of field signs – I wouldn't take a photo of every single latrine or burrow I find during a survey, but a sample of photos of such features can be helpful, particularly if you have any doubt over their identification as being from water vole. Or you might find that, at some point in the future, you need to provide evidence to someone reviewing your results.

What happens next?

Normally, you'll be doing a water vole survey for a particular reason, and a report is likely to be produced. Whether that's the case or not, it's also important to ensure that any records of water voles are widely available – you should therefore send your records to the Local Environmental Records Centre, or equivalent, as well as any other relevant local organisations (assuming that you aren't required by a contract or other agreement to keep them confidential). And you should do this as soon as possible to ensure that the information is available to those who may need it. You'll need to find out what format the information needs to be provided in before sending it, to make sure it is as 'usable' as possible. The better our collective knowledge of the distribution of water voles, the better our chances of conserving them!

Figure 11.1 Water vole feeding. © Simon Booth Photography.

Bibliography

British Standards Institution (2013) *BS42020:2013 Biodiversity – Code of Practice for Planning and Development*. BSI Standards Limited, London.

Couzens, D., Swash, A., Still, R. and Dunn, J. (2017) *Britain's Mammals: A Field Guide to the Mammals of Britain and Ireland*. Princeton University Press, Woodstock.

Dean, M. (2003) Development mitigation for water voles: a research project into the effectiveness of 'displacement' as a mitigation technique. *In Practice* 39: 10–14.

Dean, M., Strachan, R., Gow, D. and Andrews, R. (2016) *The Water Vole Mitigation Handbook*. Mammal Society Mitigation Guidance Series. Eds Fiona Mathews and Paul Chanin. Mammal Society, London.

Gelling, M., Harrington, A.L., Dean, M., Haddy, E.C., Marshall, C.E. and Macdonald, D.W. (2018) The effect of using 'displacement' to encourage the movement of water voles *Arvicola amphibius* in lowland England. *Conservation Evidence* 15: 20–25.

Gregory, C. (2016) *The Water Vole: the Story of One of Britain's Most Endangered Mammals*. Vertebrate Publishing, Sheffield.

Halford, C.M. and Moreton-Jones, K.J. (2018) DNA in the water: detecting the presence of water vole using environmental DNA (eDNA) analysis. *Mammal News* 180: 12–13.

Halford, C.M., Jones, K.J., Hill, D.J. and Schmerer, W.M. (2018) Use of environmental DNA analysis to detect the presence of water vole. *In Practice* 99: 35–39.

Harris, S. and Yalden, D.W. (2008) *Mammals of the British Isles: Handbook*, 4th edition. Mammal Society, Southampton.

Mammal Society (2018) *Britain's Mammals 2018: The Mammal Society's Guide to their Population and Conservation Status*. Eds F. Mathews, F. Coomber, J. Wright and T. Kendall. The Mammal Society, London.

Moorhouse, T.P., Gelling, M. and Macdonald, D.W. (2008) Effects of forage availability on growth and maturation rates in water voles. *Journal of Animal Ecology* 77: 1288–1295.

Moorhouse, T.P., Gelling, M. and Macdonald, D.W. (2009) Effects of habitat quality upon reintroduction success in water voles: evidence from a replicated experiment. *Biological Conservation* 142: 53–60.

Morris, P.A., Morris, M.J., MacPhearson, D., Jefferies, D.J., Strachan, R. and Woodroffe, G.L. (1998) Estimating numbers of the water vole Arvicola terrestris: a correction to the published method. *Journal of Zoology* 246: 61–62.

Ryland, K. and Kemp, B. (2009) Using field signs to identify water voles – are we getting it wrong? *In Practice* 63: 23–25.

Stewart, R.A., Jarrett, C., Scott, C., White, S.A. and McCafferty, D.J. (2018) Water vole (*Arvicola amphibius*) abundance in grassland habitats of Glasgow. *The Glasgow Naturalist* (online 2018) 27 (1). www.glasgownaturalhistory.org.uk/gn27_1/ Stewart_Glasgow_WaterVoles.pdf (accessed January 2021).

Stewart, R.A., Scott, C.M. and Macmillan, S. (2019) Interim guidelines for the conservation management of urban grassland water voles. www.glasgow.gov. uk/CHttpHandler.ashx?id=45426&p=0 (accessed January 2021).

Strachan, R. (1997) *Water Voles*. Whittet Books, London.

Strachan, R. (2010) *Mammal Detective*, 2nd edition. Whittet Books, Stowmarket.

Strachan, R. and Jefferies, D.J. (1993) The water vole *Arvicola terrestris* in Britain 1989–1990: its distribution and changing status. Vincent Wildlife Trust, London.

Strachan, C., Strachan, R. and Jefferies, D.J. (2000) Preliminary report of the changes in the water vole population of Britain as shown by the national surveys of 1989–1990 and 1996–1998. Vincent Wildlife Trust, London.

Strachan, R., Moorhouse, T. and Gelling, M. (2011) *Water Vole Conservation Handbook*, 3rd edition. Wildlife Conservation Research Unit, Oxford.

Vincent Wildlife Trust (2003) The water vole and mink survey of Britain 1996–1998 with a history of the long-term changes in the status of both species and their causes. Vincent Wildlife Trust, London.

Index